Lutz Riegger

Back-end Processing in Lab-on-a-Chip Fabrication

Lutz Riegger

Back-end Processing in Lab-on-a-Chip Fabrication

Südwestdeutscher Verlag für Hochschulschriften

Imprint

Any brand names and product names mentioned in this book are subject to trademark, brand or patent protection and are trademarks or registered trademarks of their respective holders. The use of brand names, product names, common names, trade names, product descriptions etc. even without a particular marking in this work is in no way to be construed to mean that such names may be regarded as unrestricted in respect of trademark and brand protection legislation and could thus be used by anyone.

Publisher:
Südwestdeutscher Verlag für Hochschulschriften
is a trademark of
Dodo Books Indian Ocean Ltd., member of the OmniScriptum S.R.L Publishing group
str. A.Russo 15, of. 61, Chisinau-2068, Republic of Moldova Europe
Printed at: see last page
ISBN: 978-3-8381-2089-8

Zugl. / Approved by: Freiburg, Albert-Ludwigs-Universität, Diss., 2010

Copyright © Lutz Riegger
Copyright © 2011 Dodo Books Indian Ocean Ltd., member of the OmniScriptum S.R.L Publishing group

Back-end Processing in Lab-on-a-Chip Fabrication

Dissertation zur Erlangung des Doktorgrades
der Technischen Fakultät
der Albert-Ludwigs-Universität Freiburg im Breisgau

vorgelegt von
Diplom-Ingenieur Lutz Riegger

Freiburg im Breisgau, 2010

Dekan

Prof. Dr. Hans Zappe

Referenten

Prof. Dr. Roland Zengerle

Prof. Dr. Holger Reinecke

Tag der Abgabe: 09.06.2010

Tag der Prüfung: 04.11.2010

Lutz Riegger

Lehrstuhl für Anwendungsentwicklung (Prof. Dr. Roland Zengerle)

Institut für Mikrosystemtechnik – IMTEK

Technische Fakultät

Albert-Ludwigs-Universität Freiburg

Abstract

This thesis presents back-end processing steps for polymer labs-on-a-chip which permit the realization of complex applications like biological assays on-chip. Pre-treatment for surface cleaning, hydrophilization to promote capillary action, selective hydrophobization for flow control, dry reagent pre-storage to enable fully integrated chips as well as biocompatible sealing to ensure operation and biological activity after processing are covered. Most of these processing steps are developed on a sample lab-on-a-chip which aims for the on-chip amplification of mRNA by nucleic acid sequence based amplification (NASBA).

In detail, pre-treatment of chips with hydrogen peroxide is introduced which effectively sterilizes polymer surfaces. As hydrophilic coating, PEtOx-BP, PDMAA-BP and PEG are evaluated in respect to hydrophilicity. All coatings render polymer surfaces strongly hydrophilic (water contact angle $\theta < 40°$). It is further determined that only PEG exhibits full NASBA compatibility.

A new method for the selective surface patterning of microfluidic chips with hydrophobic fluoropolymers is developed and demonstrated by the fabrication of hydrophobic valves *via* dispensing. It enables efficient optical quality control for the surface patterning due to incorporated dyes thus permitting low-cost production of highly reliable hydrophobic valves. Specifically, different dyes for fluoropolymers are investigated and two fluoropolymer-solvent-dye solutions based on fluorescent quantum dots (QD) and carbon black (CB) are presented in detail. The latter creates superhydrophobic surfaces (typical layer thickness: 2 µm) on arbitrary substrates, e.g. chips made from cyclic olefin copolymer (COC, $\theta = 157.9°$), provides very good visibility for the visual quality control in polymer labs-on-a-chip and increases burst pressures of hydrophobic valves by up to 35 %. On the sample lab-on-a-chip, the quality control in combination with Teflon-CB as coating reduces the burst pressure variability from 14.5 % down to 6.1 % compared with Teflon-coated valves.

Two approaches for the pre-storage of bioreagents, i.e. spotting and drying as well as spotting and freeze-drying are presented and evaluated. Multiple 25 nL droplets of reagents are dispensed into sample reaction chambers while preventing any contact with the adjacent walls for the spotting and drying. The concept is validated for primers and probes by successfully conducting NASBA for mRNA of three different human papilloma virus (HPV) types, even after 2.5 months of storage. For spotting and freeze-drying, reagents are dispensed on frozen substrates whereas multiple droplets form small reagent pillars. These pillars are successfully freeze-dried on-chip using a custom-developed protocol. Rehydration and positive NASBA amplification of susceptible enzymes is demonstrated after three weeks.

Biocompatible sealing is realized *via* temperature diffusion bonding and adhesive bonding. The former process is optimized by the introduction of a transparent compound foil which allows for optical read-out techniques. With the adapted process, strongly bonded test chips are produced (delamination pressure > 3 bar). A potentially negative temperature impact of the sealing process on pre-stored reagents is further excluded for a minimum distance of 180 µm between interface and reagents. Finally, it is demonstrated that PEG as hydrophilic coating can interfere with temperature diffusion bonding as residual PEG on the chip surface promotes delamination. For adhesive bonding as versatile approach for the sealing of polymer microfluidic chips, the influence of process parameters is investigated. Specifically, a process chain comprising pre-processing, adhesive transfer as well as post-processing is presented and parameter recommendations are provided. As device for adhesive transfer, a modified laminator is utilized which transfers thin layers of adhesive onto only the chip surface *via* a silicone roll. Using this device and a high temperature compatible (T_g > 100°C) epoxy adhesive, adhesive layers in the range of 2 - 4 µm can be reproducibly transferred (CV < 4 %). For best bonding results, it is recommended to provide 2.5 µm thin layers of adhesive in combination with a subsequent evacuation step at 10 mbar for 3 hours. Further, it is proposed to integrate capture channels near large, featureless areas to compensate for variations in processing and thus prevent clogging of channels. With these recommendations, strongly bonded test chips (delamination pressure > 4 bar) can be reliably produced (yield > 80 %). Additionally, the process is inherently compatible with Vistex as hydrophilic coating due to similar coupling chemistry.

Zusammenfassung

Die hier vorgestellte Arbeit beschreibt nach der Fertigung durchgeführte Prozessschritte für aus Polymeren hergestellte Substrate, mit deren Hilfe sich komplexe Applikationen wie biologische Assays auf einer Lab-on-a-Chip Plattform realisieren lassen. Abgedeckt werden Vorbehandlung zur Oberflächenreinigung, Hydrophilisierung zur Unterstützung von kapillaren Befüllungsvorgängen, selektive Hydrophobisierung zur definierten Flußkontrolle, Trockenreagenzienvorlagerung zur Realisierung von vollintegrierten Systemen sowie biokompatible Deckelung um die generelle als auch biologische Funktionalität zu ermöglichen. Die Entwicklung der meisten dieser Prozessschritte wird auf einem Lab-on-a-Chip System durchgeführt, mit welchem sich mRNA durch die sogenannte NASBA-Reaktion amplifizieren läßt.

Im Detail wird eine Vorbehandlung der Polymer-Substrate mit Wasserstoffperoxid vorgeschlagen um diese effektiv zu sterilisieren. Als hydrophile Beschichtung werden PEtOx-BP, PDMAA-BP und PEG untersucht. Mit allen diese Beschichtungen können stark hydrophile Oberflächen (Kontaktwinkel gegen Wasser $\theta < 40°$) hergestellt werden. Es wird des Weiteren gezeigt, dass nur PEG vollständige Kompatibilität zur NASBA-Reaktion aufweist.

Eine neu entwickelte Methode zur selektiven Oberflächenbeschichtung von mikrofluidischen Substraten mit hydrophoben Fluoropolymeren wird anhand der Herstellung von hydrophoben Ventilen durch Dispensieren demonstriert. Die Methode ermöglicht effiziente optische Qualitätskontrolle durch die Verwendung von Farbstoffen und damit die kostengünstige Herstellung von zuverlässigen hydrophoben Ventilen. Speziell werden verschiedene, mit Fluoropolymerlösungen mischbare Farbstoffe untersucht und zwei Polymer-Farbstoff-Lösungen, basierend auf fluoreszierenden Quantum-Dots sowie Ruß, im Detail präsentiert. Mit Hilfe von Ruß lassen sich superhydrophobe Schichten mit typischen Dicken von 2 µm auf jeglichen Oberflächen herstellen, zum Beispiel auf zyklischen-Olefin-Copolymeren (COC, $\theta = 157.9°$). Zudem verfügen die Schichten über einen sehr guten Kontrast für die visuelle Qualitätskontrolle auf polymerbasierten Lab-on-a-Chip-Substraten und verstärken die Ventile um bis zu 35 %. Auf dem Beispiel-Lab-on-a-Chip System führt die Qualitätskontrolle in Zusammenspiel mit der Teflon-Ruß Beschichtung zu einer Verbesserung der Ventilvariabilität von 14.5 % auf 6.1 % im Vergleich zu nur Teflon-beschichteten Ventilen.

Zwei Ansätze für die Vorlagerung von Reagenzien, namentlich das Dispensieren und Trocknen sowie das Dispensieren und Gefriertrocknen, werden vorgestellt und evaluiert. Bei dem ersten Ansatz werden mehrere 25 nL Tropfen von Reagenzienlösung in Reaktionskammern dispensiert wobei jeglicher Kontakt der Reagenzienlösungen mit den Kammerwänden ausgeschlossen werden kann. Der Ansatz wird validiert für Primer und Sonden durch die erfolgreiche NASBA-Amplifikation von mRNA dreier verschiedener Pappilomaviren (HPV) direkt nach der Trocknung sowie nach zweieinhalb Monaten der Lagerung. Für das Dispensieren und Gefriertrocknen werden Reagenzienlösungen auf gefrorene Substrate dispensiert. Bei der Abgabe von mehreren Tropfen frieren Folgetropfen direkt auf den vorherig dispensierten Tropfen ein. Die gefrorenen Reagenzien werden erfolgreich gefriergetrocknet mit einem speziell hierfür entwickelten Protokoll. Die erfolgreiche NASBA-Amplifikation von mRNA mit auf diese Weise gefriergetrockneten Enzymen wird nach 3 Wochen gezeigt.

Die Biokompatible Deckelung wird durch Heißlaminieren sowie eines Klebeprozesses realisiert. Der Prozess des Heißlaminierens wird hierbei durch die Einführung einer transparenten Verbundfolie optimiert, da dadurch optische Auslesetechniken eingesetzt werden können. Mit dem angepaßten Prozess können stark belastungsresistente Verbünde hergestellt werden (Delaminationsdruck > 3 bar). Ein übermäßiger Temperatureintrag für vorgelagerte Reagenzien durch den Deckelungsprozess kann hierbei für einen Abstand von mindestens 180 µm zwischen Reagenzien und Deckelunterseite ausgeschlossen werden. Abschließend wird gezeigt, dass PEG-Rückstände auf der Substratoberfläche in Zusammenspiel mit dem Heißlaminieren zur Delamination der Verbundfolie führen können. Für den Klebeprozess als vielseitiger Ansatz für die Deckelung von polymerbasierten mikrofluidischen Substraten werden Einflußparameter auf den Gesamtprozeß eingehend untersucht. Im Detail wird eine Prozeßkette von der Vorbehandlung über den Kleberauftrag bis hin zur Nachprozessierung vorgestellt, inklusive mehrerer Parameterempfehlungen. Zum Klebeauftrag wird ein modifizierter Laminator verwendet mit welchem sich definiert dünne Klebeschichten auf Substratoberflächen unter Zuhilfenahme einer Silikonwalze übertragen lassen. Ein Hochtemperaturepoxidkleber (T_g > 100°C) lässt sich hiermit im Schichtdickenbereich zwischen 2 - 4 µm reproduzierbar (CV < 4 %) auf Substrate aufbringen. Zur Erreichung bester Ergebnisse wird das Auftragen von 2.5 µm dicken Schichten in Kombination mit einem nachfolgenden Evakuierungsschritt bei 10 mbar für 3 h empfohlen. Außerdem empfiehlt sich das Integrieren von sogenannten Fängerkanälen neben größeren, unstrukturierten Flächen um Prozessvariationen und damit das Risiko der Kanalblockierung durch überschüssigen Kleber zu minimieren. Mit diesen Empfehlungen lassen sich stark belastungsresistente Verbünde (Delaminationsdruck > 4 bar) zuverlässig herstellen (Yield > 80 %). Zusätzlich ist der Prozeß durch die ähnliche Kopplungschemie kompatibel mit Vistex als hydrophiler Beschichtung.

Own Publications

Journal Articles

[1] L. Riegger, M. Grumann, T. Nann, J. Riegler, O. Ehlert, W. Bessler, K. Mittenbühler, G. Urban, L. Pastewka, T. Brenner, R. Zengerly, J. Ducrée, Read-out concepts for multiplexed bead-based fluorescence immunoassays on centrifugal microfluidic platforms, *Sensors Actuators A*, **126**, pp. 455-462, 2006.

[2] L. Riegger, M. Grumann, J. Steigert, S. Lutz, C.P. Steinert, C. Mueller, J. Viertel, O. Prucker, J. Rühe, R. Zengerle, and J. Ducree, Single-step Centrifugal Hematocrit Determination on a 10-$ Processing Device, *Biomed. Microdevices*, **9**, pp. 795-799, 2007.

[3] L. Riegger, M. M. Mielnik, A. Gulliksen, D. Mark, J. Steigert, S. Lutz, M. Clad, R. Zengerle, P. Koltay and J Hoffmann, Dye-based coatings for hydrophobic valves and their application to polymer labs-on-a-chip, *J. Micromech. Microeng.*, 20, **4**, 045021, 2010.

[4] L. Riegger, O. Strohmeier, B. Faltin, R. Zengerle and P. Koltay, Adhesive bonding of microfluidic chips: Influence of process parameters, *J. Micromech. Microeng.*, 20, **8**, 087003, 2010.

[5] M. Grumann, A. Geipel, L. Riegger, R. Zengerle and J. Ducrée, Batch-mode mixing on centrifugal microfluidic platforms, *Lab Chip*, Vol. 5, **5**, pp: 560-565, 2005.

[6] J. Steigert, M. Grumann, T. Brenner, K. Mittenbühler, T. Nann, J. Rühe, I. Moser, S. Haeberle, L. Riegger, J. Riegler, W. Bessler, R. Zengerle, and J. Ducrée, Integrated Sample Preparation, Reaction, and Detection on a High-frequency Centrifugal Microfluidic Platform, *Journal of the Association for Laboratory Automation (JALA)*, 10, **5**, pp. 331-341, 2005.

[7] M. Grumann, J. Steigert, L. Riegger, I. Moser, B. Enderle, K. Riebeseel, G. Urban, R. Zengerle, and J. Ducrée, Sensitivity Enhancement for Colorimetric Glucose Assays on Whole Blood, *Biomed. Microdevices*, **8**, **3**, pp. 209-214, 2006.

[8] J. Steigert, M. Grumann, M. Dube, W. Streule, L. Riegger, T. Brenner, P. Koltay, K. Mittmann, R. Zengerle, and J. Ducrée, Direct Hemoglobin Measurement on a Centrifugal Microfluidic Platform for Point-of-Care Diagnostics, *Sensors & Actuators A*, **130-131**, pp. 228-233, 2006.

[9] J. Steigert, M. Grumann, T. Brenner, L. Riegger, J. Harter, R. Zengerle,

and J. Ducrée, Fully Integrated Whole Blood Testing by Real-Time Absorption Measurement on a Centrifugal Platform, *Lab Chip*, **8**, pp. 1040-1044, 2006.

[10] M. Boettcher, M. S. Jaeger, L. Riegger, J. Ducrée, R. Zengerle and C. Duschl, Lab-on-chip-based cell separation by combining dielectrophoresis and centrifugation, *Biophys. Rev. Lett.*, **1**, pp. 443–51, 2006.

[11] J. Steigert, T. Brenner, M. Grumann, L. Riegger, S. Lutz, R. Zengerle, and J. Ducrée, Integrated Siphon-Based Metering and Sedimentation of Whole Blood on a Hydrophilic Lab-on-a-Disk, *Biomed. Microdevices*, **9**, pp. 675-679, 2007.

[12] L. Furuberg, M. M. Mielnik, A. Gulliksen, L. Solli, I.-R. Johansen, J. Voitel, T. Baier, L. Riegger, F. Karlsen, RNA amplification chip with parallel microchannels and droplet positioning using capillary valves, *Microsystem Technologies*, **14**, pp. 673 - 681, 2008.

[13] H. Keegan H, T. Baier, T. Hansen-Hagge, F. Karlsen, A. Gullicksen, P. Gronn, L. Solli, M. Mielnik, L. Furuberg, P. Koltay, L. Riegger, N. Bolger, J. O'Leary, C. M. Martin, Lab-on-a-Chip: The Future of Cervical Pre-Cancer Diagnostics, *Modern Pathology*, **23**, p. 96A, 2010.

Conference Proceedings

[14] L. Riegger, M. Grumann, T. Brefka, J. Steigert, C.P. Steinert, T. Brenner, R. Zengerle, and J. Ducrée, Bubble-Free Priming of Blind Capillaries for High-Accuracy Centrifugal Hematocrit Measurements, Proc. microTAS, Boston, USA, pp. 796-798, 2005.

[15] L. Riegger, J. Steigert, M. Grumann, S. Lutz, G. Olofsson, M. Khayyami, W. Bessler, K. Mittenbuehler, R. Zengerle, and J. Ducrée, Disk-Based Parallel Chemiluminescent Detection of Diagnostic Markers for Acute Myocardial Infarction, Proc. microTAS, Tokyo, Japan, pp. 819-821, 2006.

[16] L. Riegger, M. Grumann, J. Steigert, R. Zengerle, J. Ducrée, Microfluidics on a Conventional, 10-$ CDROM Drive: All-In-One Determination of the Hematocrit, Proc. microTAS, Tokyo, Japan, pp. 1555-1557, 2006.

[17] L . Riegger, J. Steigert, S. Lutz, W. Streule, R. Zengerle, and J. Ducrée, Automated Hematocrit Measurement and Patient Data Labelling by a Commercial DVD-Writer with a Low-Cost Optical Add-On, Proc. microTAS, Paris, France, pp. 1249-1251, 2007.

[18] L. Riegger, W. Streule, C. Henze, R. Zengerle and P. Koltay, PipeJet P4.5 – Kostengünstiger Kompakt-Nanoliter-Dosierer im Mikrotiterplattenformat, Proc. MST Kongress, Dresden, Germany, pp. 925-926, 2007.

[19] L. Riegger, M. M. Mielnik, D. Mark, W. Streule, M.Clad, R. Zengerle, and P. Koltay, Teflon-Carbon Black as new Material for the Hydrophobic Patterning of Polymer Labs-on-a-chip, Proc. Transducers 2009, Denver, USA, 2026-2029, 2009.

[20] M. Grumann, I. Moser, J. Steigert, L. Riegger, A. Geipel, C. Kohn, G. Urban, R. Zengerle, J. Ducrée, Optical Beam Guidance in Monolithic Polymer Chips for Miniaturized Colorimetric Assays, Proc. IEEE-MEMS, Miami, USA, pp. 108-111, 2005.

[21] M. Grumann, M. Dube, T. Brefka, J. Steigert, L. Riegger, T. Brenner, K. Mittmann, R. Zengerle, and J. Ducrée, Direct Hemoglobin Measurement by Monolithically Integrated Optical Beam Guidance, Proc. Transducers 2005, Seoul, Korea, Vol.2, 1106-1109, 2005.

[22] M. Grumann, L. Riegger, T. Nann, O. Ehlert, K. Mittenbuehler, G. Urban, L. Pastewka, T. Brenner, R. Zengerle and J. Ducrée, Parallelization of chip-based fluorescence immunoassays with quantum-dot labelled beads, Proc. Transducers, Seoul, Korea, Vol. 2, pp 1114–7, 2005.

[23] T. Brenner, M. Grumann, S. Haeberle, L. Riegger, J. Steigert, R. Zengerle, and J. Ducrée, An Extended Toolbox for Fully Integrated Sample Processing and Detection on a High-Frequency Centrifugal Microfluidic Platform, Proc. 4M Conference, Karlsruhe, Germany, pp. 413-417, 2005.

[24] J. Steigert, L. Riegger, M. Grumann, T. Brenner, J. Harter, R. Zengerle, and J. Ducrée, Rapid Alcohol Testing in Whole Blood by Disk-Based Real-Time Absorption Measurement, Proc. microTAS, Boston, USA, pp. 1-3, 2005.

[25] M. Grumann, M. Dube, O. Gutmann, S. Lutz, J. Steigert, L. Riegger, K. Mittmann, M. Daub, R. Zengerle, and J. Ducrée, Rapid Centrifugal Processing of Microarray Experiments, Proc. microTAS, Boston, USA, pp. 328-330, 2005.

[26] J. Ducrée, M. Grumann, T. Brenner, J. Steigert, L. Riegger, S. Haeberle, R. Zengerle, Bio-Disk: Eine vielseitige, zentrifugale mikrofluidische Plattform für die Point-of-Care Diagnostik, Proc. MST Kongress, Freiburg, Germany, pp. 199-203, 2005.

[27] M. Grumann, J. Steigert, L. Riegger, M. Dube, S. Lutz, T. Brenner, J. Harter, R. Zengerle, J. Ducrée, Vollintegrierte Alkoholbestimmung in Vollblut über CD-basierte, zeitaufgelöste Absorptionsmessung, Proc. of MST-Kongress, Freiburg, Germany, pp. 847-851, 2005.

[28] J. Ducrée, S. Haeberle, T. Brenner, M. Grumann, L. Riegger, J. Steigert, R. Zengerle, Scaling Laws and Prospects of Centrifugal Nanofluidics, Proc. Nanofluidics, Boekelo, Netherlands, 2005.

[29] J. Steigert, T. Brenner, M. Grumann, L. Riegger, R. Zengerle and J. Ducrée, Design and fabrication of a centrifugally driven microfluidic disk for fully integrated metabolic assays on whole blood, Proc. IEEE-MEMS, Istanbul, Turkey, pp. 418–21, 2006.

[30] S. Lutz, M. Grumann, O. Gutmann, M. Dube, L. Riegger, J. Steigert, M. Daub, R. Zengerle, J. Ducrée, Centrifugal Processor For Standard Microarray Slides, Proc. microTAS, Tokyo, Japan, pp. 329-331, 2006.

[31] S. Lutz, P. Lang, L. Riegger, W. Streule, M. Daub, P. Koltay, F. v. Stetten, R. Zengerle and J. Ducrée, Contact-Free Dispensing of Living Cells in Nanoliter Droplets, Proc. Actuator, Bremen, Germany, pp. 800-802, 2008.

[32] T. Metz, L. Riegger, C. Ziegler, R. Zengerle and P. Koltay, Pressure characteristics modelling for rapid design of capillary microfluidic systems, Proc. microTAS, San Diego, USA, pp. 1354-56, 2008.

[33] J. Hoffmann, L. Riegger, D. Mark, F. von Stetten, R. Zengerle and J. Ducrée, TIR-based dynamic liquid-level and flow-rate sensing and its application on centrifugal microfluidic platforms, Proc. IEEE-MEMS, Sorrento, Italy, pp. 539–542, 2009.

[34] A. Yusof, L. Riegger, N. Paust, A. Ernst, R. Zengerle and P. Koltay, A Non-Invasive Single Cell Dispensing Approach for 2-Dimensional Micro-Patterning, Proc. Actuator, Bremen, Germany, 2010.

Patents

[35] M. Grumann, J. Ducrée, R. Zengerle, L. Riegger, Device and Method for Determining the Volume Parts of Phases in a suspension, WO/2007/042207.

Table of Contents

1 Introduction 1
1.1 Lab-on-a-Chip 1
1.2 Back-end Processing of Labs-on-a-Chip 2
    1.2.1 Surface Modification 3
    1.2.2 Reagent Storage 5
    1.2.3 Biocompatible Sealing 6
1.3 Structure of Thesis 7

2 Fundamentals 9
2.1 Microfluidics 9
    2.1.1 Surface Tension 9
    2.1.2 Viscosity 11
    2.1.3 Contact Angle 13
    2.1.4 Capillary Pressure 15
2.2 Microactive Project 18
    2.2.1 HPV 18
    2.2.2 Screening for HPV 19
    2.2.3 NASBA 20
    2.2.4 Amplification Chip Principle & Functionality 22

3 Surface Modification 24
3.1 Chip Pre-Treatment 24
    3.1.1 Cleaning by Ultrasonication 24
    3.1.2 Hydrogen Peroxide Treatment 24
    3.1.3 Plasma Treatment of Chip Surface 27
3.2 Hydrophilic Coating 27
    3.2.1 Means of Application 27
    3.2.2 Coating Evaluation 29
    3.2.3 NASBA Compatibility 32
    3.2.4 Processing Compatibility and Mid-Term Stability 33
    3.2.5 PCR-Compatible Hydrophilic Coating 34
3.3 Hydrophobic Patterning 35
    3.3.1 Solvents for Fluoropolymers 35
    3.3.2 Dyes for Fluoropolymer Solutions 36
    3.3.3 Chip Patterning 42
3.4 Conclusion & Outlook 52

4 Reagent Pre-Storage 53
4.1 Principle & Volume Calibration 53
4.2 Reagent Spotting and Drying 56
    4.2.1 Spotting into Reaction Chambers 57
    4.2.2 NASBA Results 60

4.3	Reagent Spotting & Freeze-Drying		61
	4.3.1	Spotting on Frozen Substrate	61
	4.3.2	Freeze-Drying	63
	4.3.3	NASBA Results	67
4.4	Reagent Bead Fabrication		67
4.5	Conclusions & Outlook		69

5 Biocompatible Sealing   70

5.1	Temperature Diffusion Bonding		70
	5.1.1	Principle	70
	5.1.2	Materials and Methods	71
	5.1.3	Evaluation	74
5.2	Adhesive Bonding		81
	5.2.1	Principle	81
	5.2.2	Materials and Methods	82
	5.2.3	Evaluation	85
5.3	Conclusion & Outlook		94

6 General Conclusions and Outlook   95

7 References   97

7.1	Literature	97
7.2	Used Materials	109
7.3	Used Devices	110

8 Appendix   112

8.1	Process Recommendation		112
	8.1.1	Low Temperature Applications	112
	8.1.2	High Temperature Applications	114

9 Nomenclature   115

9.1	List of Symbols	115
9.2	List of Abbreviations	116

Chapter 1
Introduction

1.1 Lab-on-a-Chip

Medical diagnostic devices are utilized in hospitals, laboratories, at the doctor's office and in near-patient or point-of-care environments [1]. For centralized diagnostics, tests are carried out in large laboratories. Except for emergency situations, it is not likely for patients to receive examination results on the same day since economical constraints require the collection of samples prior to testing to obtain larger batches [2]. This major drawback can be circumvented by single-test devices for near-patient environments [3] also referred to as point-of-care devices. These devices are typically developed as desktop or portable systems [4] thus the results can be provided much faster. Further, tests can be performed at standard hospital laboratories, practitioners offices or at home [5].

Consequently, a major interest in commercial and scientific development has been the transfer of applications from centralized diagnostics to near-patient and point-of-care environments [5,6,7,8,9,10,11,12]. In this approach, microfluidics play a dominant role [13,14,15,16] since the required miniaturization of medical tests relies on the accurate and precise handling of fluids in microscale dimensions. Near-patient or point-of-care applications additionally require full process integration, reduced consumption of sample and reagents, reduced size of the device as well as short times-to-result [5].

So-called lab-on-a-chip (LoaC) or micro total analysis systems (µTAS) based on microfluidic technologies meet these special requirements [17,18,19,20,21,22,23,24]. They typically comprise a reusable detection and actuation unit and a disposable microfluidic chip. So far, key laboratory unit operations such as sample preparation [25], separation [26], metering [26], mixing [27], reaction and detection [28] have been successfully demonstrated by separate stand-alone lab-on-a-chip systems. Additionally, various complex tasks like simultaneous DNA amplification and detection [29], HIV immunoassay and detection [30], single-cell processing [31], and pathogen-specific DNA extraction from whole blood [32] have been realized. Most relevant for point-of-care environments however are compact, fully integrated systems which are capable of running autonomously [33,34,35,36]. A system capable of sample preparation as well as PCR on-chip has been reported by D. Chen et al. [36] (figure 1.1). It features liquid as well as dry reagent storage combined with conventional fluidic elements like mixers and valves. The downside of the concept however is the overabundance of active components (e.g., valves) and chip-to-actuation-unit interfaces.

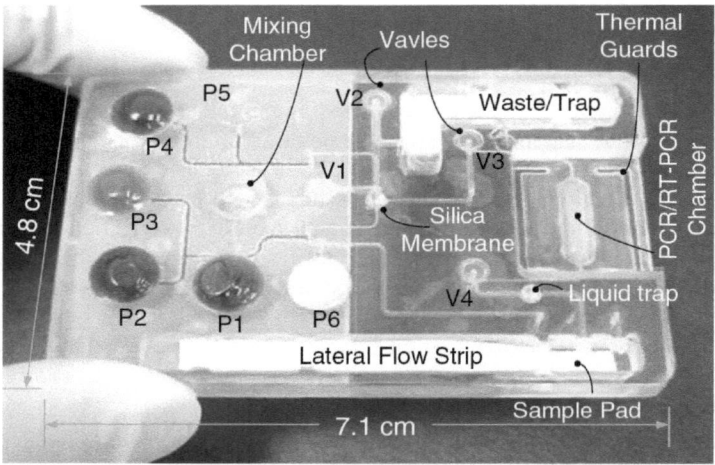

Figure 1.1 Fully integrated lab-on-a-chip for the conduction of PCR, including sample preparation (courtesy of D. Chen [36]).

Several lab-on-a-chip systems are well competitive to conventional technologies [6,37]. Thus, they comply with the technical and economic market requirements. This qualifies LoaC technology as the forerunner in medical diagnostics for near-patient and point-of-care environments. However, the wide-scale commercial success will be determined by the ability to meet the stringent end-user demands on reliability, user friendliness and most importantly the costs per test.

1.2 Back-end Processing of Labs-on-a-Chip

When focusing on disposable microfluidic chips as part of lab-on-a-chip systems, further referred to as labs-on-a-chip, one can differentiate between silicon/glass-based or polymer-based labs-on-a-chip. The former are typically processed in clean-room environments using existing infrastructure from microelectronics and MEMS. Although clean-room processed chips can offer small pricing in the case of mass production, the cost per chip increases drastically with the used area. Due to the relatively large area necessary for microfluidic channels and the expenses for clean-room process development, a fabrication in e.g silicon is not necessarily advantageous in the case of disposable, low-cost microfluidic chips. Thus, there has been a strong trend towards polymer labs-on-a-chip [12,23,24] specifically due to their

1.2 Back-end Processing of Labs-on-a-Chip

amenability for low-cost mass-production.

Polymer labs-on-a-chip are not inherently capable of realizing complex applications like biological assays without further back-end processing steps after fabrication. Further, in economic terms, these steps can make up to 80% of the manufacturing cost [38]. For transporting liquids, the chips at least have to be sealed to be able to operate. Additionally, the surface of the chip may have to be modified to promote capillarity and provide flow control. Further, a fully integrated chip for biological applications demands the pre-storage of reagents. Processing steps to provide typical functionality for complex labs-on-a-chip are schematically shown in figure 1.2.

In this work, relevant processing steps are categorized in the following way: surface modification comprises the cleaning as well as pre-treatment like plasma activation, the application of hydrophilic surface coatings and hydrophobic patterning, i.e. a selective application of fluoropolymers. Reagent storage describes the pre-storage of bioreagents on-chip whereas the biocompatible sealing encompasses methods for the bonding of microfluidic chips while ensuring biological functionality after processing.

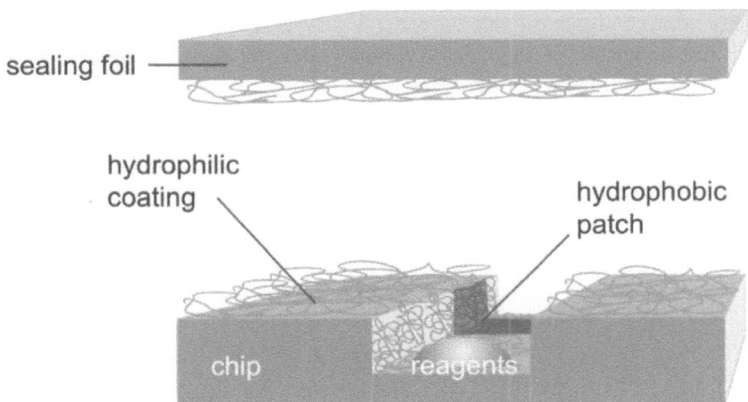

Figure 1.2 Chip model to demonstrate potential processing steps, i.e. hydrophilic coating, hydrophobic patterning, reagent pre-storage and sealing.

1.2.1 Surface Modification

During prototyping, polymer microfluidic chips are typically processed in a non-sterile and non-clean room environment. Therefore, it is necessary to develop steps as to ensure that debris and dust particles are removed and the risk of contamination is minimized. After these cleaning steps, the chip may

1.2 Back-end Processing of Labs-on-a-Chip

have to be activated by e.g. a plasma process [39] to provide a better wetting surface for subsequent coatings [40]. In biological applications, a hydrophilic coating to prevent unspecific absorption [41,42] is typically used. The thus modified surface further facilitates priming of channels *via* capillary action.

To conduct complex protocols of unit operations, the fluid flow has to be controlled in a defined manner. This can be realized by the integration of passive valves based on hydrophilic coatings. A simple way for retaining liquids is the use of hydrophobic patterns for flow control due to the change in surface tension [43,44,45]. Another approach utilizes capillary valves, i.e. geometrical restrictions to pin liquid plugs at defined positions due to a pressure drop [46,47]. Both concepts can be combined in the form of hydrophobic valves [48,49], i.e. capillary valves featuring a hydrophobic coating. They provide higher retention forces than the purely capillary valves and exhibit a better reliability as fabrication tolerances are compensated by the coating. A sophisticated microfluidic chip featuring multiple hydrophobic valves to enable subsequent execution of unit operations has been reported by Per Andersson et al. [49,50] (figure 1.3). Here, the hydrophobic coating is applied by aligning a plastic or metal shadow mask over the chip and spraying a solution of fluoropolymers over the mask.This method allows for the fabrication of highly reliable valves ($CV < 1$ %) however the coating principle has the downside of high dead volumes and the requirement for multiple masks, depending on the geometry.

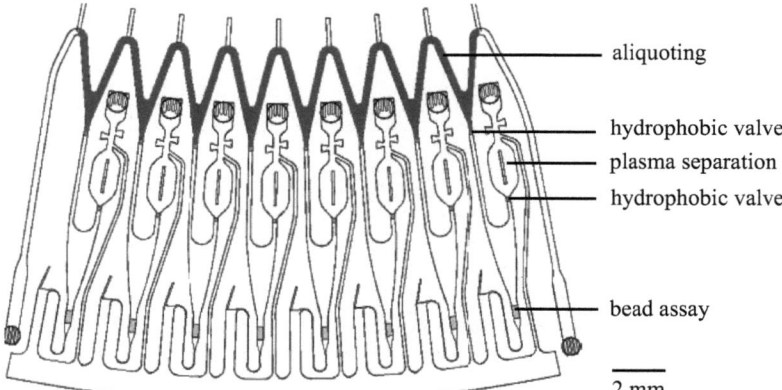

Figure 1.3 Principle drawing of the Gyrolab Bioaffy. The microfluidic disk features structures for aliquoting, plasma separation, bead assays as well as hydrophobic valves for flow control (courtesy of P. Andersson [49]).

1.2.2 Reagent Storage

A fully integrated lab-on-a-chip capable of conducting complex bioassays is required "to have everything on-board" which primarily refers to the pre-storage of reagents. Possible approaches are liquid reagent storage or dry reagent storage. As most bioreagents are subject to a variety of chemical reactions in aqueous solutions and tend to loose activity or denaturate over time [51], dry reagent storage is highly preferable. Especially for the long-term storage of bioreagents, lyophilized reagents in the form of powder or beads fabricated by freeze-drying are primarily used [51]. This allows for the dehydration of sensitive material like proteins and can thus largely increase the shelf-life of reagents.

For macroscale experiments, commercially produced reagent beads (diameter ~ 5 mm) are readily available. These however can not be applied in compact labs-on-a-chip due to space respectively total volume constraints. One approach is the direct freeze-drying of reagents in reaction chambers which has been reported by M. Brivio et al. [52] (figure 1.4). Here, reaction volumes down to 6 µL containing enzymes were pipetted into reaction chambers, frozen, freeze-dried and subsequently reactivated on-chip. However, this concept is not viable for sub-µL reaction volumes as well as complex fluidic structures as a locally defined distribution of reagents can not be guaranteed.

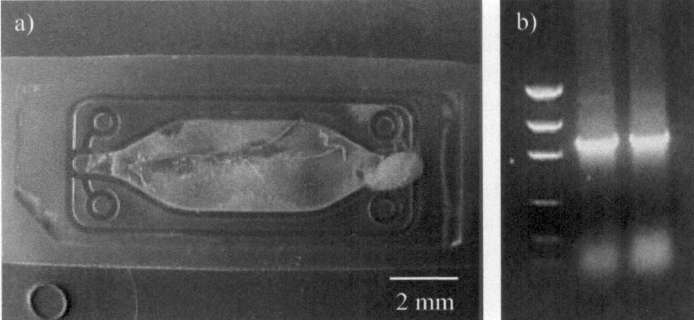

Figure 1.4 a) Image of microfluidic chip with freeze-dried reagents. b) Positive PCR gel result based on freeze-dried reagents (courtesy of M. Brivio et al. [52]).

1.2.3 Biocompatible Sealing

Bonding of polymer substrates can prove to be highly challenging and thus ultimately decide on the functioning or even potential commercialization of a lab-on-a-chip [53]. Further, the process must not inherently destroy (e.g., due to too high temperatures) or inactivate bioreagents pre-stored on-chip (i.e., being biocompatible). The simplest way for bonding of polymer chips is the use of self-adhesive foils [54,55]. However, when in direct contact with liquids, the adhesive tends to diffuse into the sample over time thus posing the risk of e.g. reaction inhibition. To counter this disadvantage, foils featuring a pressure-sensitive adhesive stored in microcapsules are used as the adhesive is only released on the chip surface. The downside of using these foils is the low bond strength and the incompatibility to hydrophilic coatings. Therefore, alternate approaches for bonding of microfluidic chips like laser welding [56], solvent bonding [57], thermal bonding [58, 59] or adhesive bonding [56,60,61,62, 63,64,65] have to be taken into account. Depending on the boundary conditions, e.g. material used, hydrophilic coating required, different approaches are more or less feasible. Laser welding for example requires an absorber at the interface between chip and lid [56] whereas solvent bonding has the drawback that a number of residue solvent bonds can remain after the sealing process [64]. Providing a universal approach for different applications is a non-trivial task but still highly desirable. Bonding via adhesives primarily relies on the wetting of the chip surface by the adhesive and is therefore applicable on various polymer materials and material combinations. The main challenge for adhesive bonding is the prevention of channel clogging by adhesive. An approach to address this issue has been presented by C. Lu et al. [64] (figure 1.5). It is based on the interstitial flow of a low viscous adhesive due to capillary action between the bonding partners. A clogging of channels is prevented for microchannels > 60 µm. Still, the bonding of isolated features based on this technique will prove difficult. Further, only low-viscous adhesives can be applied.

Figure 1.5 Cut of a microfluidic channel sealed by adhesive bonding. No channel deformation as well as channel clogging is observable (courtesy of C. Lu et al. [64]).

1.3 Structure of Thesis

The main objective of this thesis is the development of back-end processes for polymer labs-on-a-chip, i.e. pre-processing, hydrophilic surface coating, hydrophobic patterning, reagent storage as well as biocompatible sealing. Most work on lab-on-a-chip systems primarily focuses on the chip design or the application however without suitable processing, complex labs-on-a-chip can not be realized. Additionally, potential commercialization depends on the stability of processing as well as scalability and ease-of-use.

To realize this objective, existing methods are evaluated in respect to their applicability on a sample lab-on-a-chip which screens for human papilloma virus (HPV) by detecting mRNA on-chip. Additionally, novel methods which enable downscaling of existing concepts or enhance the current state-of-the-art are described. In detail, the thesis is structured as follows:

1.3 Structure of Thesis

2	Fundamentals	A theoretical background for the fluidic effects relevant within this thesis is given. Additionally, the reference lab-on-a-chip on which most of the development has been conducted is presented.
3	Surface Modification	The pre-treatment as well as hydrophilic surface coating is described in this section. Further, special emphasis is given on the hydrophobic patterning, i.e. the selective coating with fluoropolymers.
4	Reagent Storage	In this section, the reagent storage based on spotting and drying as well as spotting and freeze-drying is presented.
5	Biocompatible Sealing	Two methods for the biocompatible sealing of microfluidic chips, namely the temperature diffusion bonding and the adhesive bonding are described and evaluated
6	General Conclusions and Outlook	The thesis is summarized and a general outlook is given.
7	References	All references, used materials and devices of the thesis are listed.
8	Appendix	Process recommendations for low-temperature as well as high-temperature applications are provided.
9	Nomenclature	All used symbols, variables and abbreviations are listed

Chapter 2
Fundamentals

The first part of the chapter describes fundamental microfluidic effects and phenomena relevant for this thesis. Due to the large surface to volume ratio in microfluidics, surface effects and consequently capillary effects are dominating. As most work in this thesis is based on the handling of miniscule amounts of liquids, a deeper understanding of these effects is imperative.

In the second part of this chapter, a reference lab-on-a-chip on which most of the development has been conducted will be presented. As the required processing is closely linked to the biological application, a background on HPV mRNA as target together with the underlying means of detection will be given. This is followed by a detailed description of the chip functionality.

2.1 Microfluidics

2.1.1 Surface Tension

Amongst the most important phenomena in nature, especially in biology [66] are interfacial effects. The three states of matter which characterize the interfaces are solid, liquid and gas. Surface tension is a physical phenomenon at the interface between two different phases caused by intermolecular forces, primarily chemical bonds between liquid molecules which compel the liquid to form a thin surface membrane. A molecule at the surface cannot form as many bonds thus it is more difficult to move a submerged object past the surface than to move it in the liquid [67].

Surface tension is the force acting in the surface of a liquid, tending to minimize the area on the surface. Molecules at the liquid surface are subject to an inward force of molecular attraction perpendicular to the surface but experience interactions with molecules of a different phase at the interface. If the force of these interactions is lower than the inward force, the molecules at the surface will be pulled inward. The cohesive forces acting in a liquid are shown in figure 2.1.

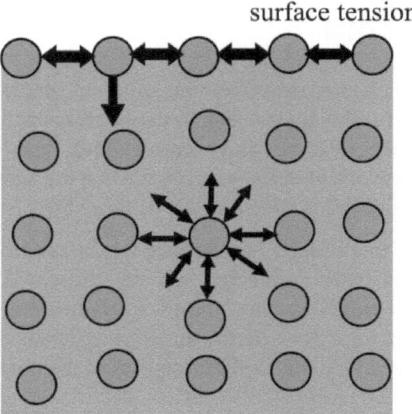

Figure 2.1 Cohesive forces between molecules in a liquid resulting in surface tension.

Surface tension can be measured with a liquid membrane enclosed in a rectangular wire frame exhibiting one movable side of length l. A force F has to be applied perpendicular to this wire in order to avoid contraction of the membrane (figure 2.2).

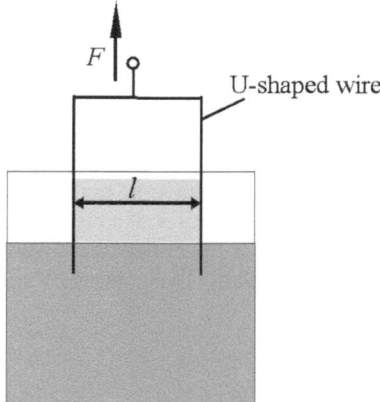

Figure 2.2 Experimental setup to measure the surface tension of a liquid.

2.1 Microfluidics

In more detail, F is the cumulative sum of individual intermolecular forces F_s at the two surfaces of the membrane These intermolecular forces F_s are generally directed tangentially to the surface, i.e. into the plane of the membrane. The surface tension is given by the force F divided by the length l of the membrane, further divided by two because the liquid wets both sides of the wire which leads to equation 2.1.

$$\sigma = \frac{F}{2l} \qquad (2.1)$$

In order to enlarge the surface area dA of a liquid, molecules from the inside of the liquid are transported to its surface which requires the energy dW. Thus, the interface area dA and the energy dW needed to form this interface are proportional and related by σ, the surface tension, resulting in equation 2.2. Inversely, the process gains energy and thus the surface contains energy, so called surface energy.

$$dW = \sigma dA \qquad (2.2)$$

2.1.2 Viscosity

Viscosity is the fluid resistance to shear or flow and is a measure of the adhesive/cohesive or frictional liquid property. It is caused by intermolecular friction exerted when layers of fluids attempt to slide by one another. The resistance of a liquid to deform under shear rate τ is characterized by the dynamic viscosity η which describes the relocatability of molecules towards each other.

A common method for measuring the viscosity of liquids is depicted in figure 2.3. Two plates of the same surface area A are placed at a distance y to each other into a homogeneous liquid. By moving the upper plate at a constant velocity u, the liquid experiences a constant shear caused by the friction between the boundary plates and the liquid. The required force F moving the upper plate defines the dynamic viscosity or simply viscosity of the liquid.

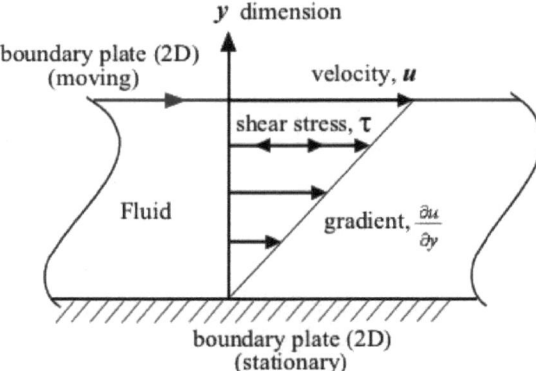

Figure 2.3 Laminar shear of a liquid encased between two parallel plates for viscosity measurement (courtesy of H. Padleckas).

Unlike a solid, a liquid deforms constantly under shear stress. The applied force F is proportional to the area A and velocity u of the plate and inversely proportional to the distance y between the plates (equation 2.3).

$$F = \eta \cdot A \cdot \left(\frac{\partial u}{\partial y}\right) \qquad (2.3)$$

Newton's theory postulates that the shear stress between the layers is proportional to the velocity gradient directed perpendicular to the layers. Liquids which satisfy this criteria are known as Newtonian fluids, e.g. water, which show an independency of their viscosity to the applied velocity. For non-Newtonian fluids, the viscosity can either increase or decrease according to the changes in shear rate, also called shear thickening or thinning. The change in viscosity over time under a constant shear rate is known as thixotropy [68].

Changes in temperature affect the viscosity of Newtonian liquids as well as non-Newtonian liquids. When the liquid is heated, the cohesive forces between molecules are reduced which eventually reduces the viscosity of the liquids. The viscosity of water in respect to temperature is depicted in figure 2.4.

2.1 Microfluidics

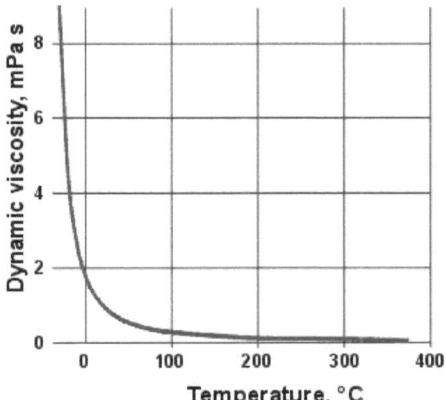

Figure 2.4 Dependency between viscosity of water and temperature T (courtesy of M. Chaplin, London South Bank University). Values for temperatures below zero degree apply for supercooled water.

2.1.3 Contact Angle

A direct measure for wetting of surfaces by a certain liquid is commonly considered to be the contact angle θ. Geometrically, it is given by the angle enclosed between the solid surface and the tangent of the liquid–vapor boundary at the three-phase contact point of the liquid (l), solid (s) and gaseous (g) phases (figure 2.5). All interfacial forces act tangentially to the surfaces of mutual phase contact.

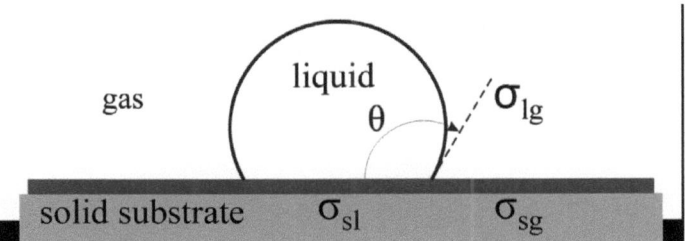

Figure 2.5 Schematic drawing of a droplet wetting a solid interface to visualize contact angle and surface tension.

A system consisting of a liquid droplet surrounded by gas placed on a solid substrate is defined by the surface tension between liquid and gas σ_{lg}, liquid and solid σ_{ls} and solid and gas σ_{sg}. The Young equation (equation 2.4) describes all these surface tension interactions in equilibrium state [69]. The contact angle is the angle formed by solid/liquid and liquid/gas interfaces.

2.1 Microfluidics

$$\sigma_{sg} - \sigma_{sl} = \sigma_{lg} \cos\theta \qquad (2.4)$$

If a liquid is dispensed on a solid material, different effects may occur which are dependent on the wettability of a surface (figure 2.6). For contact angles $\theta > 90°$, cohesive forces are dominating over adhesive forces and the liquid minimizes its contact to the so-called *hydrophobic* surface. By contrast, *hydrophilic* surfaces exhibit contact angles $\theta < 90°$ and exhibit partial or very good wetting behaviors.

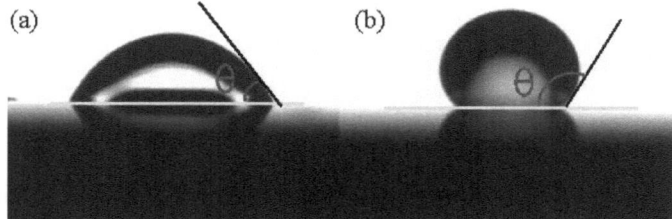

Figure 2.6 Wetting behavior of liquids. (a) Wetting droplet with a contact angle $\theta < 90°$, (b) sparsely-wetting droplet with a contact angle $\theta > 90$

Strongly hydrophobic surfaces with contact angles $\theta > 150°$ are referred to as *superhydrophobic* [70,71]. Drops that come in contact with a superhydrophobic material retain a nearly spherical shape (figure 2.7).

Figure 2.7 DI-water droplet on a superhydrophobic surface exhibiting a contact angle of $\theta > 150°$.

Superhydrophobicity can be explained by the chemical composition of a surface (i.e., low surface energy) in combination with it's surface roughness [71]. The increase in hydrophobicity arises from an increase in surface area that has to be wetted by a liquid. Above a critical value of roughness, the drop contacts only with a small fraction of the surface, trapping air between the drop and the surface [72].

2.1.4 Capillary Pressure

Another effect related to the contact angle and surface tension of a liquid is the capillarity in narrow channels or tubes. This phenomenon can be explained by considering two opposing forces: adhesion and cohesion [73]. If the adhesive intermolecular forces between the liquid and a solid are stronger than the cohesive intermolecular force, the liquid forms a concave meniscus (i.e., the interface between liquid and gas) for $\theta < 90°$. The resulting force F_{res} which acts in the capillary works against gravity and pulls the liquid upward as depicted in figure 2.8-a. For the opposite case ($\theta > 90°$.), a convex meniscus is formed and the force works in conjunction with gravity (figure 2.8-b).

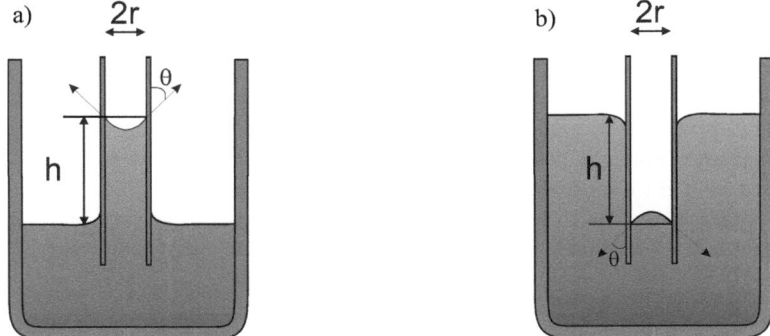

Figure 2.8 Effect of surface tension in a capillary. A concave meniscus is formed for wetting liquids (a) and a convex meniscus for non-wetting fluids (b).

Channels with inner diameter < 2 mm are referred to as capillaries. Here, surface effects typically dominate over bulk forces like gravity. The pressure difference between phases like water and air is defined as the *capillary pressure* and plays an important role in the description of liquid flow in channels. The capillary pressure in a rectangular channel of width w and height h is a function of contact angle θ and surface tension σ (equation 2.5).

$$p_\theta = 2 \cdot \sigma \cdot \cos\theta \cdot \left(\frac{1}{w} + \frac{1}{h}\right) \qquad (2.5)$$

2.1 Microfluidics

The capillary pressure is proportional to the surface tension and inversely proportional to the effective radius of the interface. In microfluidic systems, the lid often exhibits a different contact angle than the channel walls which leads to equation 2.6.

$$\Delta p_\theta = 2 \cdot \sigma \left[\cos\theta_{channel} \cdot \left(\frac{1}{w} + \frac{2}{h}\right) - \cos\theta_{lid} \cdot \frac{1}{w} \right] \qquad (2.6)$$

A difference in capillary pressure is utilized for passive valves in microfluidic systems (figure 2.9). If a liquid propagates along a microfluidic channel with height/width w_1, h_1 and reaches a step and/or a wider channel (w_2, h_2), it will stop due to a pressure drop (equation 2.7). This is referred to as capillary valve.

$$\Delta p_\theta = p_{\theta 2} - p_{\theta 1}$$
$$\Delta p_\theta = 2 \cdot \sigma \cdot \cos\theta \left[\left(\frac{1}{w_2} + \frac{1}{h_2}\right) - \left(\frac{1}{w_1} + \frac{1}{h_1}\right) \right] \qquad (2.7)$$

If the smaller channel (or restriction) exhibits a hydrophobic coating, the liquid will stop before the restriction. This is referred to as hydrophobic valve. The difference in capillary pressure at both sides of a liquid droplet in a channel is given by equation 2.8 [48]. If the liquid stops at a certain position due to a difference in capillary pressure, the respective structure can be referred to as valve. Accordingly, the required pressure to pass this valve structure which equals the pressure difference is called burst pressure.

$$\Delta p_\theta = p_{\theta 2} - p_{\theta 1}$$
$$\Delta p_\theta = 2 \cdot \sigma \left[\left(\cos\theta_2 \cdot \left(\frac{1}{w_2} + \frac{1}{h_2}\right)\right) - \left(\cos\theta_1 \cdot \left(\frac{1}{w_1} + \frac{1}{h_1}\right)\right) \right] \qquad (2.8)$$

2.1 Microfluidics

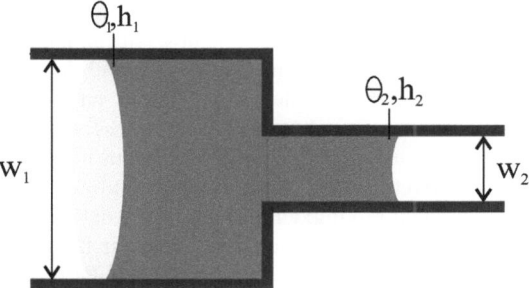

Figure 2.9 Schematic to visualize capillary pressure of a liquid plug in a channel which depends on channel widths w, channel height h and contact angle θ.

2.2 Microactive Project

The EU-project Microactive [74] aims for the development of a lab-on-a-chip system to carry out automatic, accurate diagnosis at the local doctor's office with a low cost-per-test of approximately 30 €. Here, determining the risk potential for cervical cancer by detecting human papillomavirus (HPV) mRNA transcripts from a cervical smear sample to assess genetic activity of the virus has been the main target. The project will thus integrate all necessary unit operations for detecting mRNA: cell concentration, lysis, purification, sample collection, reagent mixing, amplification and detection. This includes the development of disposable polymer chips for use in the instrument as well as a technology for storing all needed reagents on these chips. Further, reliable techniques for multi-channel parallel detection which are crucial for recognizing the disease signatures are required. In this project, two microfluidic chips are being developed, one for the sample preparation, one for conducting the amplification reaction. The work of this thesis will focus on the so-called amplification chip.

The project consortium includes three research institutes, IMTEK, SINTEF [75] and IMM [76], two companies, NorChip [77] and BioFluidix [78] who have extensive experience with mRNA and microfluidics, respectively, and are motivated to commercialize results. The project is further assisted by the Coombe Women's Hospital [79] acting as the end user, validating system usability and comparing the achieved results with clinical tests.

2.2.1 HPV

Worldwide, cervical cancer is the second most common cancer affecting women with the majority of cases occurring in developing countries [80]. The primary cause of cervical cancer is the infection with certain types of HPV [81,82]. HPV are non-enveloped and ubiquitous viruses while the viral capsid is highly resistant to the environment. Transmission occurs by direct person-to-person contact and indirectly by smear infection. More than 100 HPV genotypes infect different body areas. About 20 HPV-types infect the anogenital tract of men and women, predominantly by sexual transmission [83]. More than 90% of genital warts are caused by HPV type 6 and 11 [84]. HPV-types that cause warts are called "low-risk" HPV, whereas infections with "high-risk" HPV may result in asymptomatic cervical, vaginal and vulval intrapithelial neoplasia (CIN, VaIN and VAIN) [85], i.e. cancer precursors, which may progress to cancerous lesions within 10 to 15 years after infection [86]. However, the immune system is able to eliminate more than 90% of all high- and low-risk HPV infections. More than 99% of all cervical cancers are caused by high-risk HPV while about 70% of cervical carcinomas are caused by HPV 16 and HPV 18. Cervical carcinoma is the second most prevalent female cancer worldwide with 288000 deaths annually [87].

2.2.2 Screening for HPV

HPV has become a very important topic for women during recent years. More than half of all German women younger than 55 regularly take part in cervical cancer screening. Conventional screening is performed by cytological smears but the sensitivity of cytology is only 54% in routine screening [88]. Figure 2.10 depicts the required steps for cytological smear testing. However, cytology and colposcopy results are subjective and depend on experience of the pathologist and gynecologist. The approach to develop an automated diagnostic system has therefore been an aim for many diagnostic companies. HPV DNA detection tests, e.g. Hybrid Capture 2™ (HC2) [89], have been introduced into the market more than 10 years ago. HPV DNA tests exhibit more than 95% sensitivity for the detection of CIN, but the specificity of HPV DNA testing is unsatisfactory and considerably lower than for cytology. In most women, high-risk HPV infection of the cervix does not cause development of CIN before it is cleared by the immune system. Thus, detection of HPV DNA is no indication for treatment unless CIN is detected by colposcopy.

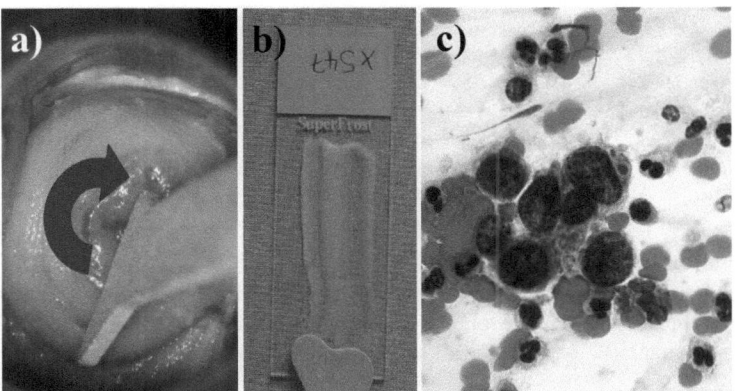

Figure 2.10 Screening for HPV via cytological smear. a) Collecting sample from cervix, b) transfer onto glass slide, c) microscopic view of CIN 3 (blue dyed cells are HPV infected).

Furthermore, the mere knowledge of being infected with high-risk HPV represents an enormous psychological burden for each patient and may lead to unnecessary colposcopy, cytology and biopsies. In addition, the cost of these unnecessary diagnostics as well as treatment measures is a considerable financial burden for public health systems worldwide. Therefore, HPV tests should rather just detect precancerous lesions, i.e. CIN 2 and especially CIN 3, which are an indication for treatment. Figure 2.11 [90] displays results of the most commonly used HPV DNA test, HC2, compared with the HPV mRNA test PreTect HPV-Proofer™ [77] which will be utilized in the Microactive project. The PreTect HPV-Proofer™ detects HPV E6/E7 mRNA (relevant for the development of pre-cancerous and cancerous lesions [91]) that is found in high quantities in CIN 2 and 3 and is absent in HPV infections without CIN lesions. Therefore, HPV E6/E7 mRNA tests are more specific for the detection of CIN 2 and CIN 3 lesions than HPV DNA tests, i.e. HPV mRNA tests show less false positive results compared with HPV DNA tests [92]. The HPV-Proofer is still the only assay for the detection of HPV E6/E7 mRNA.

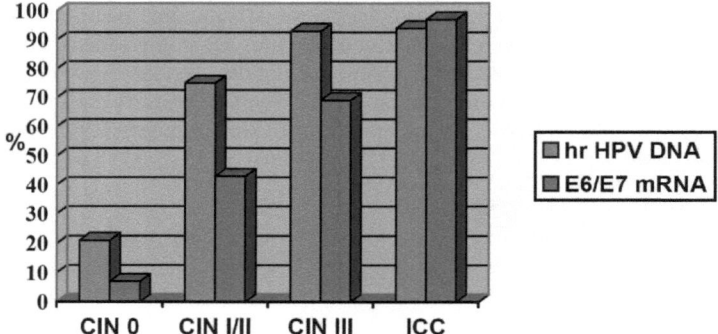

Figure 2.11 Sensitivity of Hybrid Capture 2™ test compared with PreTect HPVProofer™. The HPV DNA test gives significantly more positives in benign and low-grade lesions.

The PreTect-Proofer™ detects E6/E7 mRNA of 5 out of more than 13 HPV high-risk types. However, these 5 types (HPV 16, 18, 31, 33, 45) are the predominant HPV types found in high-grade CIN and compromise 97% of the oncogenic HPV types detected in cervical carcinomas [93].

2.2.3 NASBA

The PreTect-Proofer™ takes advantage of Nucleic Acid Sequence Based Amplification (NASBA) for mRNA detection [94]. Although RNA can also be amplified by PCR using reverse transcriptase (in order to synthesize a complementary DNA strand as a template), NASBA works isothermal (41°C) and therefore does not require an expensive thermocycler which facilitates miniaturization. However, due to the large number of different reagents (e.g., enzymes, nucleotides and primers) required for NASBA, the assay is quite susceptible to variations in back-end processing (RNases, pre-stored reagents) as well assay conduction (contamination, skill of operator). NASBA is a two-step process: an initial enzymatic amplification of the nucleic acid targets followed by detection of the amplicons. The amplicons are single-stranded and do not require an additional denaturation step prior to detection. This allows for real-time read-out of NASBA reactions. Briefly, NASBA works as follows (figure 2.12):

1. RNA template is added to the reaction mixture. Primer 1 attaches to its complementary site at the 3' end of the template.
2. Reverse transcriptase synthesizes the anti-sense complementary DNA strand.
3. RNase H destroys the RNA template (RNase H only destroys RNA in RNA-DNA hybrids, but not single-stranded RNA).
4. Primer 2 attaches to the 5' end of the synthesized anti-sense DNA strand.
5. T7 RNA polymerase produces a complementary RNA strand which serves as a new template in step 1, starting a new round of amplification.

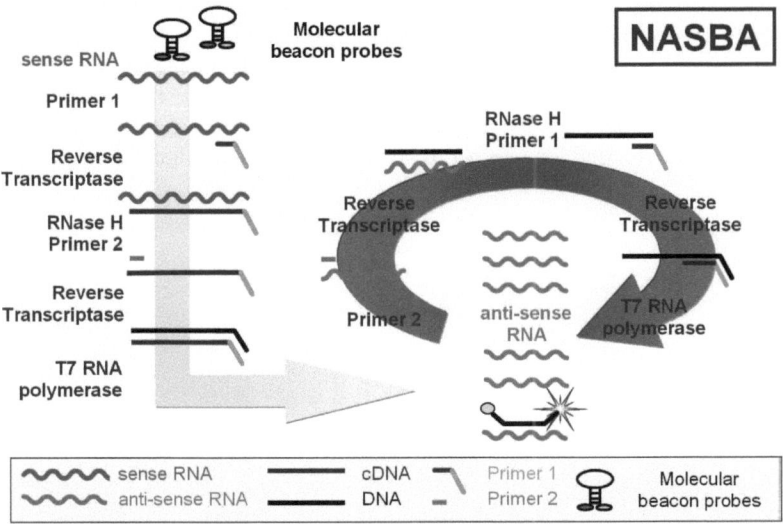

Figure 2.12 Principle of NASBA (courtesy of Norchip). The reaction comprises two different primers, molecular beacon probes as well as three enzymes.

2.2.4 Amplification Chip Principle & Functionality

For conduction of NASBA on-chip, a tight control over the stoichiometric reagent ratio is required. Thus, the sample has to be precisely metered. To promote subsequent annealing of primers during NASBA, the mastermix comprising primers and probes has to be heated to 65°C for 5 minutes which denatures secondary and tertiary RNA structures [95]. This is followed by mixing with enzymes and the actual NASBA reaction at 41°C. To be able to screen for different HPV targets, the chip has to have multiple channels, each featuring a different set of primers.

The Microactive NASBA chip (referred to as amplification chip) is fabricated by injection molding from cyclic olefin copolymer (COC) [97,M1]. It comprises eight parallel channels with three stop positions, i.e. hydrophobic valves. The three hydrophobic valves separate structures for metering, mastermix dissolution, enzyme dissolution and detection whereas the last two steps are conducted in the same chamber. The functionality of the chip is depicted in figure 2.13 (chip not to scale). The sample is dispensed into the inlet and primes the metering channels (each containing aliquots of 640 nL) by capillary action and stops at the first hydrophobic valve (figure 2.13-a). When the sample reaches the waste chamber at the end of the supply channel, the capillary pump (i.e., a sponge paper) pulls the remaining sample from the supply channel and the sample is cut off at the beginning of the metering channels [96] thus leading to exact volume definition in each channel (figure 2.13-b). Then, the sample is pulled into the reaction channels *via* syringe pump where it rehydrates the mastermix (figure 2.13-c). Next, the same step is repeated and the enzymes are rehydrated. This is finally followed by the detection after a defined number of amplification cycles (figure 2.13-d).

2.2 Microactive Project

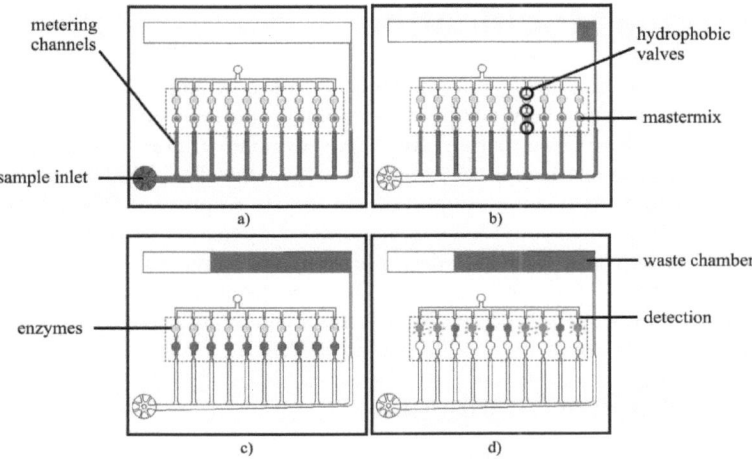

Figure 2.13 Principle of chip function: a) Sample introduction and aliquoting, b) metering, c) dissolution of mastermix, d) dissolution of enzymes and detection.

Chapter 3
Surface Modification

Typical fabrication methods for polymer microfluidic chips are CNC-micro-machining, hot embossing or injection molding. Independent of fabrication, the chips at least have to be cleaned (e.g. to remove lubricants or release agents) prior to application. For sensitive assays, the chips additionally have to be free of contaminations to prevent any negative impact on the biological reaction. To provide better wetting properties for potential coating, a surface activation step may be required. These steps are summarized as chip pre-treatment in the following paragraphs. The subsequent section summarizes the evaluation of different hydrophilic coatings which is followed by presenting the process development for hydrophobic patterning.

3.1 Chip Pre-Treatment

3.1.1 Cleaning by Ultrasonication

The first step in chip processing is the removal of residual debris or lubricants from the microfluidic channels by a 30' ultrasonication step in DI-water or 2-propanol, respectively, followed by drying with pressurized nitrogen.

3.1.2 Hydrogen Peroxide Treatment

The treatment with hydrogen peroxide is a common way to sterilize materials [98]. Specifically, it is highly suitable to denaturate RNases and DNases which possibly prevent the detection of targets in lowly concentrated samples. Storing substrates in 3% H_2O_2 for 24 hours before further processing guarantees the complete denaturation of RNases and DNases on the substrate surface. Removing the hydrogen peroxide solution by rinsing with RNase-free water [M2] minimizes the risk of recontamination. After rinsing, each substrate should be dried with pressurized nitrogen and stored in a sterile container.

Testing for RNase Contamination

The amplification chip acts as a detection platform for mRNA transcripts. RNases are present on the human skin as to protect from virus RNA. As a consequence, they can be found in large abundance in any laboratory. For tests based on detecting mRNA, it is imperative to avoid RNase contamination. However, due to the sensibility of RNA-based tests in standard lab environments, the screening for RNase contamination is a viable approach to qualify the back-end processing in respect to biological applications.

3.1 Chip Pre-Treatment

This paragraph describes a quick way of testing for RNase contamination based on the RNaseAlert™ kit [M3]. This assay uses a ribo-oligonucleotide comprising a fluorophore and quencher. The oligonucleotide does not fluoresce until degraded by RNase as the fluorophore is separated from the quencher. The RNaseAlert™ Kit detects a variety of nucleases, including RNase A and RNase T1 and RNase 1. To conduct the assay, samples are mixed with the test solution and incubated at 37°C for one hour. The assay can also be applied to microfluidic chips by priming the channels with the test solution and incubating it accordingly. The results can help to identify weak spots in the chip processing chain, i.e. which steps are most susceptible to RNase contamination. The test protocol is as follows:

- Mix 45 µL of sample with 5 µL of RNaseAlert™ buffer with lyophilized fluorescent substrate
- Prime sealed chip with test solution
- Incubate test solution at 37°C for 60 min
- Acquire fluorescence images utilizing blue excitation
- Evaluate mean grey value of channels in fluorescence images, subtract background fluorescence, compare with negative control (NC). Samples exhibiting signals two times higher than NC signals are considered RNase-contaminated

Representative fluorescence images [D1,D2] of test chambers are depicted in figure 3.1. Here, the fluorescent signal of the positive control is about 60 times higher than the negative control confirming that the test has been conducted successfully.

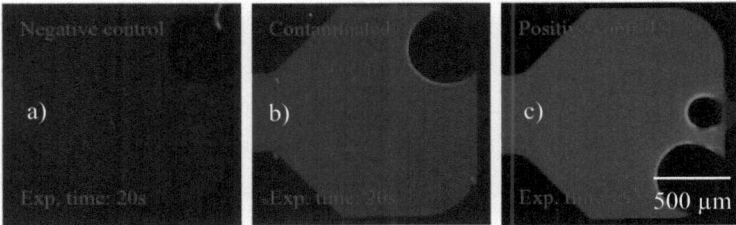

Figure 3.1 Fluorescence images of test chips after a) H_2O_2 cleaning, b) DI water cleaning and c) RNaseA contamination (positive control). The H_2O_2 treated chip features the lowest fluorescence while the DI-water cleaned chip expresses a small RNase contamination per definition. Note that for a) and b), the exposition time is 10 times longer.

3.1 Chip Pre-Treatment

The test results are summarized in figure 3.2. It can be seen that untreated chips are considered RNase contaminated. The increase of contamination due to ultrasonication in DI-water is supposedly caused by RNases present in the water. Therefore, the contamination is at its highest after the ultrasonication step which expresses a ratio of 3.6 compared to the negative control. The treatment with H_2O_2 is an effective step to destroy RNase on the chip surface – H_2O_2 treated chips presumably contain even less RNase than the RNaseAlert™ test tubes. The process steps of plasma activation (please refer to section 3.1.3) and PEG (please refer to section 3.2) coating exhibit an increase in relative contamination, due to e.g. transport of the chips, but the effect is comparable for all channels and does not resolve in too high a contamination. The hydrophobic patterning (please refer to section 3.3) exhibits the highest variation in fluorescence signal after the H_2O_2 treatment but the mean signal is still below the contamination cut-off. This can be explained as the chips are exposed to the most sub-steps (move the chips to the holder, pattern the chips, wipe the top surface, clean the chips with pressurized nitrogen) of all the back-end processing steps thus increasing the risk of single channel contamination. However, the presented hydrophobic patterning procedure should still provide acceptable results for RNA-based tests.

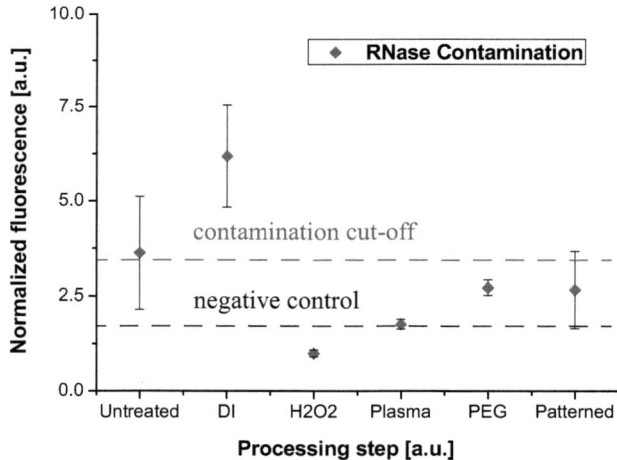

Figure 3.2 Normalized mean fluorescence intensity of processed test chips for quantifying possible RNase contamination. At least six reaction chambers of the respective chip have been measured. Untreated as well as DI-water cleaned chips express contamination while chips tested after subsequent back-end processing are considered non-contaminated.

3.1.3 Plasma Treatment of Chip Surface

Prior to coating, the polymer surface is activated with oxygen plasma to clean the chip surface and to provide hydroxylation. Before plasma irradiation, e.g. a COP surface exhibits a hydrogen atom bonded with tertiary carbon [97]. The plasma irradiation breaks the molecular bonds and removes hydrogen atoms. Thus, a radical carbon is created which can easily be bonded with oxygen, hydrogen or hydroxyl groups [99]. This process results in a temporarily hydrophilized surface, promoting a better surface wetting for possible coatings. The plasma activation is conducted in a standard plasma reactor [D3] while utilizing the following parameters:

- Evacuate plasma chamber $p = 1$ Pa
- Ingest oxygen gas $p = 10$ Pa
- Microwave generator 120 s at $P = 200$ W

Activated surfaces keep their hydrophilic properties for a limited time only, depending on the process parameters. To the author's experience, further coatings should be applied within the same week.

3.2 Hydrophilic Coating

Untreated COC surfaces exhibit a water contact angle of ~ 92° [12] and are therefore considered hydrophobic. For chips utilizing self-priming structures *via* capillary action, a long-term stable hydrophilization is required. Furthermore, stored bioreagents typically include proteins which are likely to be adsorbed on channel surfaces of the chip. With suitable hydrophilic coatings, the protein adsorption can be suppressed [41,42]. Additionally, due to the high surface to volume ratio in microchannels, the surface roughness can lead to reaction inhibition [100]. Consequently, a hydrophilic coating can also be applied to planarize microchannels.

3.2.1 Means of Application

Dip-Coating

For applying hydrophilic coatings, on non-planar surfaces, dip-coating is considered highly suitable as the layer thickness is quite defined (presumably except for the channel edges) due to the typically automated processing. In dip-coating (figure 3.3), the substrate is attached to an upward and downward moving rod. By moving downward, the substrate is brought into a polymer solution and later extracted with a defined upward velocity. The layer formation is based on the deposition of molecules from the solution to the adsorbing substrate surface by evaporation of the solvent during extraction.

3.2 Hydrophilic Coating

The layer thickness of the resulting polymer film not only depends on the amount of the adherent and the concentration of the polymer solution, but also on the extraction velocity of the chip. For all dip-coating experiments [D5], the latter is set to 1 mm/s which has proven to result in ~ 40 nm thick layers for 5 wt% polymer-alcohol solutions [101].

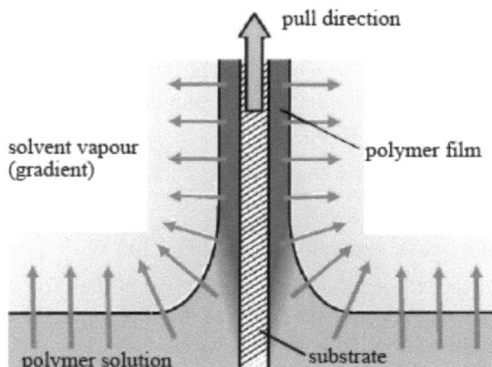

Figure 3.3 Schematic representation of the dip-coating process. The substrate is extracted from a polymer solution. The segregated polymer film on the substrate is formed from the adhering liquid by evaporation.

Via dip-coating, the complete chip including the top surface is covered by the coating solution. This can pose a problem for later sealing of the microfluidic chip. Thus, a another approach is also followed which is discussed in the following paragraph.

Pipetting

In this approach, the coating solution is directly pipetted into microfluidic channels. As each chip has to be processed manually and the meniscus of the liquid coating solution stops at arbitrary positions, the layer thickness is quite undefined. Therefore, dispensing [D4] of polymer solution into channels has been evaluated. In general, this would enable the automated processing of the chips so it should feature a better layer thickness reproducibility. Due to the very difficult handling of the alcohol-based polymer solutions (wetting of the nozzle tip, deviation of the droplet trajectory over time), development for this approach has been stalled as the manual pipetting can provide acceptable results (i.e., complete priming of chip). Still, a spill on the chip surface can not be completely prevented with this method. An example coating protocol for the amplification chip (please refer to section 2.2.4) is denoted in the following:

3.2 Hydrophilic Coating

- Slowly dispense 15 µL of coating solution into inlet
- Dispense 15 µL of coating solution into outlet/waste chamber
- Dispense 1.5 µL of coating solution into reaction chambers
- Slowly dispense 15 µL of coating solution into inlet reservoir

3.2.2 Coating Evaluation

To characterize the wetting behavior of the different coatings, namely PDMAA-BP [41], PEtOx-BP [41,102], and PEG [M4], the static contact angle of deionized water is measured [D6] (sessile drop method) on the modified substrate surfaces at six different positions on five substrates. The results for the substrate- to- substrate variation are presented in figure 3.4. The substrates have been coated *via* dip-coating in Methanol containing 5 wt% of the respective coating. The PDMAA-BP and PEtOx-BP coated chips are subsequently irradiated with UV light for one hour (365 nm, 2 mW/cm^2). Here, the benzophenone group reacts due to radical coupling with an alkyl group of adjacent polymers to from a C-C bond [102].

Plain COC is expected to be slightly hydrophobic [12] which is confirmed by the measurements. It expresses a mean water contact angle of 91° with a *CV* of 3.8%. All the coatings change the surface property to strongly hydrophilic, i.e. 33.3° with a *CV* of 6.2 % for PDMAA-BP, 39,1° with a *CV* of 30.3 % for PEtOx-BP, and 32.6° with a *CV* of 21.3 % for PEG.

3.2 Hydrophilic Coating

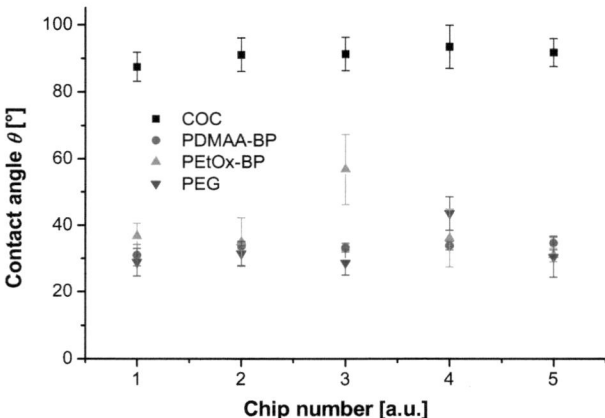

Figure 3.4 Contact angles of plain COC substrates and modified surfaces. All coatings alter the contact angle from slightly hydrophobic to strongly hydrophilic. PDMAA-BP expresses the highest reproducibility of the tested coatings.

The deviating values for the PEtOx-BP coating can possibly be explained by degenerated stock used for the experiments. PEG, in general, shows the best wetting behavior although the coupling of the layer to the surface is mainly based on adhesion in contrast to the covalent bonding for PDMAA-BP and PEtOx-BP.

Next, Teflon AF [M5] is dispensed *via* standard pipette on the substrate surface and the water contact angle is determined on these spots (figure 3.5). All coatings exhibit about the same wetting behavior after hydrophobization, i.e. 113.6° with a *CV* of 1.4 % for PDMAA-BP, 113,4° with a *CV* of 1.3 % for PEtOx-BP, and 113.6° with a *CV* of 3.2 % for PEG. The lower mean contact angle for plain COC of 104° with a *CV* of 3.7 % can be explained by the hydrophobic nature of the COC and thus a reduced wetting of the Teflon solution in contrast to the other coatings.

3.2 Hydrophilic Coating

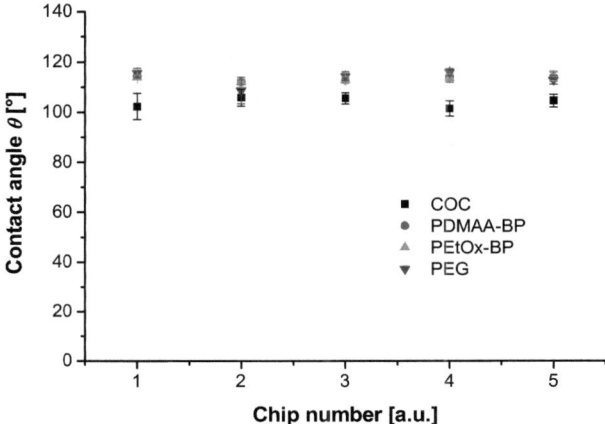

Figure 3.5 Contact angles of Teflon AF coated hydrophobic spots on plain and coated COC chips. The spots on all coated chips exhibit about the same wetting behavior.

Measuring the water contact angle on coatings gives a good reference about the general hydrophilicity respectively hydrophobicity. However, for most applications, samples containing detergents or solvents are utilized. As a consequence, the contact angles of typical PCR samples, i.e. sorbitol as stabilizer [103], DMSO to promote DNA denaturation [104] and a reference enzyme mix as well as mastermix have been measured (table 3.1). It can be seen that both selected hydrophilic coatings provide sufficiently low contact angles to provide priming by capillary forces. Further, it should be noted that the enzyme mix would not be retained by a Teflon-coated valve structure.

Table 3.1 Contact angle of NASBA-relevant media on different surface coatings. Both hydrophilic coatings, namely PEG and PDMAA-BP, feature roughly the same wetting properties.

Coating / Media	Sorbitol q / CV	45 % DMSO q / CV	Enzymes q / CV	Mastermix q / CV
COC	96.6° 3.8 %	55.1° 3.6 %	26.6° 16.2 %	76.3° 5.7 %
PEG	56.7° 18.0 %	33.7° 9.3 %	full wetting	32.6° 10.9 %
PDMAA-BP	48.9° 16.2 %	23.4° 14.9 %	full wetting	31.8° 7.0 %
COC Teflon AF	106.9° 3.1 %	83.3° 4.0 %	61.8° 6.7 %	103.1° 4.5 %
PEG Teflon AF	109.1° 5.4 %	97.3° 4.1 %	65.2° 3.6 %	106.3° 1.8 %
PDMAA-BP Teflon AF	112.1° 3.6 %	93.3° 4.8 %	63.2° 6.6 %	104.9° 3.9 %

3.2.3 NASBA Compatibility

Besides the wetting properties, the assay compatibility of the hydrophilic coatings is of prime importance for labs-on-a-chip. All coatings have been previously evaluated in respect to unspecific protein adsorption [101]. However, experiments have shown that only PEG as coating is suitable for NASBA based on cervical carcinoma (CaSki) cell line samples (table 3.2) and consequently also clinical samples. Thus, PEG is selected as hydrophilic coating for the amplification chip.

Table 3.2 NASBA-compatibility for selected hydrophilic coatings. Only with PEG as coating, all CaSki cell line samples could be amplified (courtesy of Norchip).

Coating	0.1 μM HPV 16 oligo positive / # runs	CaSki 1:1 positive / # runs
PDMAA-BP	10 / 10	2 / 10
PEtOx-BP	10 / 10	6 / 10
PEG	10 / 10	10 / 10

3.2 Hydrophilic Coating

3.2.4 Processing Compatibility and Mid-Term Stability

In order to verify that PEG does not lose its hydrophilicity during processing, water contact angles are measured after two critical processing steps (please refer to section 4.3.2). The results displayed in figure 3.6 exhibit no major loss in hydrophilicity due to storage at -80°C and freeze-drying.

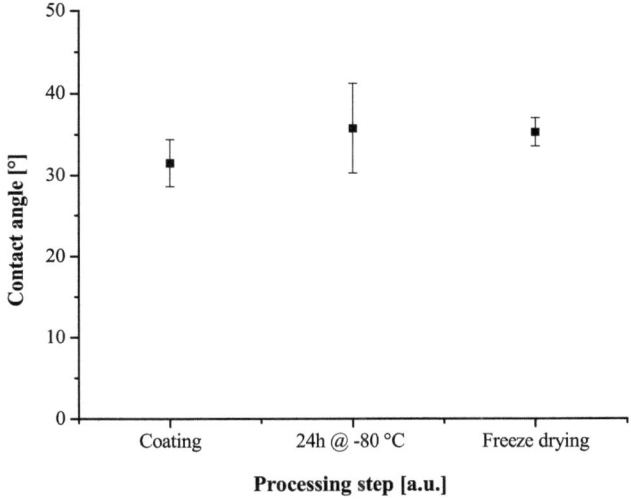

Figure 3.6 Contact angles of dip-coated and dispense-coated chips throughout processing.

For measuring the mid-term stability (figure 3.7), 80 chips are coated with PEG *via* dispensing and dip-coating. Half of the chips are previously plasma activated which results in four different batches of 20 coated chips each. Between weekly frequented measurements, the chips are stored in sterile containers. A distinct difference between dip-coated and pipette-coated chips is observable. This can be explained by the omission of the plasma activation step. The thinner coating applied *via* dip-coating (~ 40 nm [101]) does not seem to sufficiently wet the surface of the chip presumably due to it's roughness unless the latter surface is activated, thus the higher contact angle. The here pipetted layers of PEG exceed one micron in thickness thus completely covering the chip surface. Chips with pipette-coated PEG surfaces show an excellent mid-term stability of the layer while the impact of previous plasma activation is negligible. For dip-coated chips, the plasma activation reduces the variation of hydrophilicity over the chip surface and improves the mid-term stability of the layers.

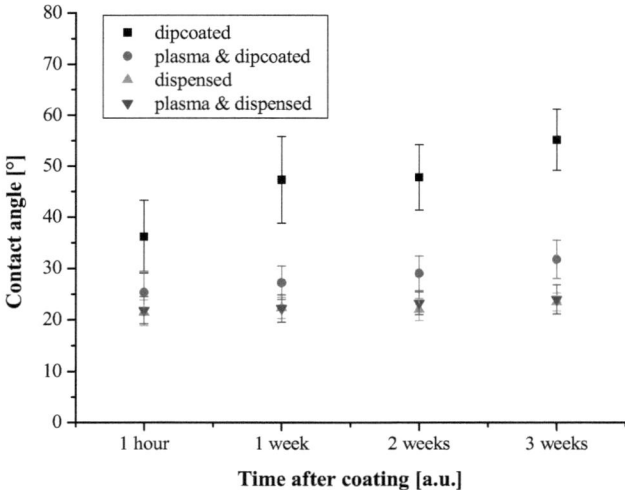

Figure 3.7 Contact angle measurements of coated substrates at different time intervals to evaluate the mid-term stability of the coating.

3.2.5 PCR-Compatible Hydrophilic Coating

The so far evaluated coatings can be readily applied for low-temperature applications like immunoassays or NASBA (PEG only). However, for conducting PCR, a high-temperature coating (glass transition temperature $T_g > 100°C$) has to be applied. One suitable candidate is Vistex [M6,106]. Recommended constitution is 2 wt% Vistex in 2-propanol with a surface coverage > 50 nL/mm^2 which ensures planarization of particularly rough surfaces (e.g., CNC-micromachined channels). The coated chips have to be cured for 45 min at a temperature of at least 120°C to ensure crosslinking of the coating and covalent coupling to the polymer substrate.

3.3 Hydrophobic Patterning

Hydrophobic patterning refers to the selective (in contrast to overall) coating of surfaces with fluoropolymers. Predominantly, it is used for the fabrication of hydrophobic valves, i.e. capillary valves featuring a hydrophobic coating. The coatings can be applied by various means, e.g. pipetting, the use of a felt pen [107], plasma processes [108,109], spray coating [49,110] or dispensing [111,112].

The typically sub-micron thick hydrophobic coatings exhibit an insufficient visibility especially in transparent polymer labs-on-a-chip. As measuring each burst pressure (i.e., the required pressure to move a liquid plug past the valve, please refer to section 2.1.4) for different patterning parameter settings is an inefficient approach, enabling a rapid visual inspection with the use of dyes is a highly preferable method for quality control. Further, depending on the dimensions of the capillary valve to be coated, a tight control of the applied volume is required to minimize the risk of overflow or insufficient coverage. Thus, different means of application are more or less suitable depending on the boundary conditions.

In the following paragraphs, fluoropolymer-solvent-dye solutions enabling the visual inspection of the localized coating and with it a quality control for the hydrophobic patterning are presented. The novel Teflon-carbon black (CB) solution allows for the creation of superhydrophobic surfaces on arbitrary substrates and thus the fabrication of stronger hydrophobic valves. The favorite coating candidates are then applied via nL-dispensing on the amplification chip. Finally, the valve coating is quantified by measuring the burst pressures.

3.3.1 Solvents for Fluoropolymers

It has been evaluated which solvents for fluoropolymers [113] are more suitable in respect to handling and applicability in hydrophobic patterning. The selected solvents (table 3.3) feature similar surface tensions (13 mN/m – 18.5 mN/m), different viscosities (0.4 mPas – 6.4 mPas) and strongly varying vapor pressures (432 – 88600 Pa) [114,115,116,117].

The test solutions are based on 0.5 wt%Teflon AF [M5] as reference fluoropolymer. For some solvents, the fluoropolymer solution has to be filtered due to insufficient solvability. The different solutions are then dispensed into valve structures (please refer to section 3.3.3) and it is observed if any overflow occurs. As a result, different viscosities and surface tensions only have a negligible impact on the spotting results. For liquids featuring a lower vapor pressure however, there is a major increase in overflow as the menisci of the fluoropolymer solution are able to creep along the channel walls for a long time before evaporating. From these results, having a very high vapor pressure would be preferable (e.g., FREON-11 as solvent) although the handling gets increasingly difficult due to fast solvent evaporation and thus potential clogging of dispenser nozzles. Thus, a solvent with a mediocre vapor pressure is deemed most suitable for application (i.e., Fluorinert FC-77 [M7]).

Table 3.3 Overview of potential solvents for fluoropolymers. All solvents exhibit a very low surface tension.

Company	Product	Surface Tension (mN/m)	Viscosity (mPas)	Vapor Pressure (Pa)	Density (g/cm^3)
3M	FC 77	13	1.3	5600	1.78
3M	FC 40	16	3.4	432	1.83
3M	HFE 7200	13.6	0.61	30000	1.43
Sigma	FREON-11	17	0.42	88600	1.49
Sigma	Perfluoro-1,3-dimethyl cyclohexane	16.6	1.9	13000	1.83
Sigma	Perfluoro-decaline	17.6	5.1	1170	1.94

3.3.2 Dyes for Fluoropolymer Solutions

Due to the fluorinated or perfluorinated nature of solvents used for dissolving fluoropolymers, standard polar dyes (chromophores and fluorophores) as well chromophores featuring non-polar groups are inapplicable. This can be explained by the highly non-polar nature of the solvents and by the absence of dipolar forces as well as the formation of hydrogen bonds in the solution. One possibility is the use of fluorinated azomethine dyes which have been synthesized by the group of S. Schrader [118,119]. Different derivatives of this type of dye have been mixed with the solvents containing fluoropolymers. For one derivative (type T 42), a good solubility (figure 3.8) and a similar hydrophobicity (~ 120°) compared to the non-dyed fluoropolymer solution has been measured although the visibility in a microchannel is insufficient (i.e., low contrast in respect to the substrate). Another possibility for dying of a fluoropolymer solution is the use of a microemulsion based on Reichardt's dye and acetonitrile [120] (figure 3.8). Again, the dried spots exhibit similar hydrophobicity but the visibility is insufficient (please refer to table 3.5).

3.3 Hydrophobic Patterning

Figure 3.8 Picture of azomethine dye (left) and Reichardt's dye (right) solved with Teflon AF in Fluorinert FC-77.

Commercially available fluorescent quantum dots (QDs) [121] are not soluble in perfluorinated solvents. Still, QDs [M8] disperse if mixed with different fluorinated solvents. The quality of the dispersion is not very reproducible and there is a strong variation in relative fluorescence intensity for different types of QDs (i.e., emission wavelengths) [M9]. Thus, the solubility has been tested in different fluorinated solvents and a fluorochlorocarbon (Freon-11 [M10]) has been found suitable for this application. The downside of using the fluoropolymer-QD-Freon solution is the difficult handling due to its very high volatility (i.e., its fast evaporation). Thus, alternate possibilities for dying the fluoropolymer solution have been researched. Carbon black [M11] (CB) has been identified as a fourth possible dye which in combination with Teflon is mainly used in fuel cells. If ultrasonicated in a fluoropolymer solution, a stable dispersion is created. Upon drying and evaporation of the solvent, a fluoropolymer-carbon black layer is formed comprising carbon black particles with a size of roughly 50 nm encased in the fluoropolymer. This not only provides a very good visibility but also superhydrophobic properties (contact angle > 150°) if a high enough particle density (i.e., total applied volume per surface area) is deposited (table 3.4).

3.3 Hydrophobic Patterning

Table 3.4 Water contact angles of Teflon AF and Teflon-carbon black coatings (featuring a similar surface coverage) on a COC surface. For the coating with low carbon black content (0.1 wt%), the surface does not exhibit superhydrophobic behavior and the contact angle is in the range of a solely Teflon-coated surface.

Surface Coverage	Teflon AF & Carbon Black Content in Solution	Contact Angle
100 nL/mm²	0.5 wt% Teflon AF	119.5° +/- 0.8°
100 nL/mm²	0.5 wt% Teflon AF 0.10 wt% carbon black	124.4° +/- 2.1°
100 nL/mm²	0.5 wt% Teflon AF 0.125 wt% carbon black	157.8° +/- 3°
100 nL/mm²	0.5 wt% Teflon AF 0.175 wt% carbon black	159.0° +/- 2.6°
100 nL/mm²	0.5 wt% Teflon AF 0.250 wt% carbon black	158.8° +/- 2.5°
100 nL/mm²	0.5 wt% Teflon AF 0.50 wt% carbon black	156.1° +/- 3.5°°

The effect can be explained by the formation of a micro-nano binary structure [110,122]. This provides the required surface roughness [70,71,72] to create superhydrophobic surfaces together with low surface energy polymer coatings (please refer to section 2.1.3). An image showing a 2 µL droplet marginally wetting a Teflon-CB surface (silicon wafer substrate) is shown in figure 3.10.

3.3 Hydrophobic Patterning

Figure 3.9 DI-water droplet on a Teflon-CB coated surface. (Bottom right) Image of water droplet acquired during contact angle measurement.

In figure 3.10, an atomic force microscopy (AFM) measurement [D7] of a Teflon-CB surface (table 5.2 solution 5) is depicted. It exhibits a distinct surface roughness with a roughness average of 134 nm. In comparison, injection-molded valve structures of the amplification chip exhibit a roughness average of 38 ± 6 nm (measured *via* white light interferometry [D8], not shown).

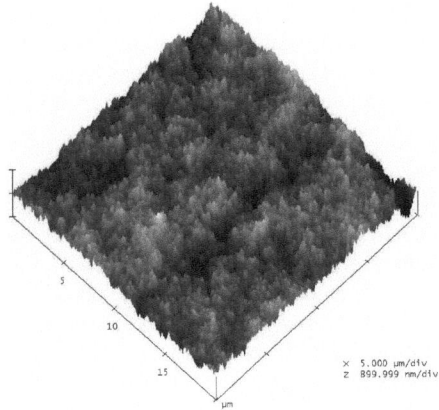

Figure 3.10 AFM image of a Teflon-CB surface (100 nL/mm^2, 0.25wt% carbon black) exhibiting a roughness average of 134 nm.

Further, Teflon-CB coated substrates are investigated with the use of a scanning electron microscope (SEM) [D9]. An example image (figure 3.11) displays a partially porous, rough as well as fine-grained topography.

3.3 Hydrophobic Patterning

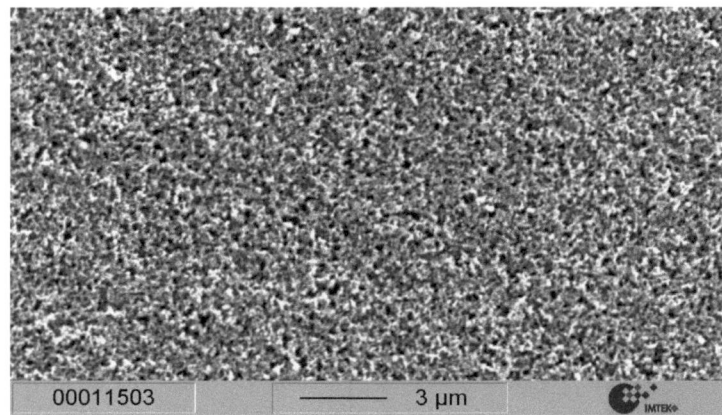

Figure 3.11 SEM image of a Teflon-CB surface (100 nL/mm^2, 0.5wt% carbon black) which exhibits a distinctively rough surface.

To summarize, different fluoropolymer-dye solutions based on a 0.5 wt% solution of Teflon AF have been investigated for contrast and hydrophobicity. The respective solution compositions, static contact angles as well as relative contrast are summarized in table 3.5. The contrast has been determined by measuring the visible light transmission of a coated glass slide (100 nL/mm² coverage). As devices, a standard microscope [D10] and CCD camera [D2] have been utilized. The mean grey scale values have been normalized to the highest absorbing sample to result in relative contrast values.

When analyzing the results, the solution based on Reichardt's dye exhibits similar hydrophobicity compared to the Teflon-only coating. However, it does not feature good contrast (i.e., << 50 %) to enable visual quality control. Teflon-QD coated surfaces provide sufficient fluorescence when compared to non-coated surfaces (SNR > 10 for the given setup) and can thus be easily discriminated. Still, this comes at the cost of reduced hydrophobicity. Differently diluted Teflon-CB solutions all feature good contrast (> 50 %). Additionally, Teflon-CB coated surfaces are considered superhydrophobic for high surface coverage (please refer to table 3.4).

3.3 Hydrophobic Patterning

Table 3.5 Table presenting six fluoropolymer solutions, the respective water contact angles and relative contrast of the coating on a transparent surface (*Note: for the Teflon-QD surface, no contrast value but a signal to noise ratio SNR is given due to its use as a fluorescent dye. A mean SNR > 10 has been measured for the valves with a standard fluorescence microscope and CCD camera). With CB as dye, superhydrophobic surfaces featuring high contrast can be created.

#	Fluoropolymer & Dye Content	Solvent	Contact Angle	Relative Contrast	Image
1	0.5 wt% Teflon AF	FC-77	119.5° +/- 0.8°	0 %	
2	0.5 wt% Teflon AF 1.0 wt% Reichardt's Dye	FC-77 2.0 wt% Acetonitrile	125.0° +/- 1.6°	7 %	
3	0.5 wt% Teflon AF 0.025 wt% QD	Freon-11	107.6° +/- 2.5°	SNR > 10*	
4	0.5 wt% Teflon AF 0.125 wt% CB	FC-77	157.8° +/- 3.0°	87 %	
5	0.5 wt% Teflon AF 0.25 wt% CB	FC-77	157.9° +/- 2.1°	98 %	
6	0.5 wt% Teflon AF 0.5 wt% CB	FC-77	156.1° +/- 3.5°	100 %	

3.3.3 Chip Patterning

Chip Layout & Processing

As proof-of-principle, the amplification chip (figure 3.12-a, please refer to section 2.2.4) is patterned with the presented fluoropolymer solutions. The chip features three hydrophobic valves (figure 3.12-b) per channel, valve 1 for a defined metering of the sample, valve 2 for confining the sample during the rehydration of dried mastermix and valve 3 for the sample confinement in the amplification chamber during read-out. The consecutive valves exhibit increasing burst pressures due to decreasing width and depth of the structures thus permitting sequential transfer of the sample plugs by increasing the pressure level. On the side of each valve, an overflow reservoir is integrated which was intentionally intended as a side-feed structure. This concept however did not result in acceptable coating results. Specifically, the coating solution propagates randomly along the restriction when dispensed in the reservoir which requires droplet volumes > 50 nL to ensure good coverage. This is tantamount to a high risk of clogging. Still, the reservoir acts as a compensation structure for overflow during spotting which is why it has not been removed from the chip design.

To permit capillary priming of the channels, the chip is coated by pipetting 60 µL of a 5 wt% PEG-methanol solution into the channels. Further, the complete chip is sealed with a polyolefin foil featuring a pressure-sensitive, silicone-based adhesive [M12]. It should be noted that a low-quality sealing (i.e., sagging of the sealing foil into the restrictions) can have a strong impact on the burst pressures and should thus be prevented by e.g. using a soft stamp to release the pressure-sensitive adhesive. Before priming the channels by capillary action, the inlet as well as the outlets are opened with the use of a scalpel. A syringe pump for plug movement is connected via an adapter plate to the outlets.

3.3 Hydrophobic Patterning

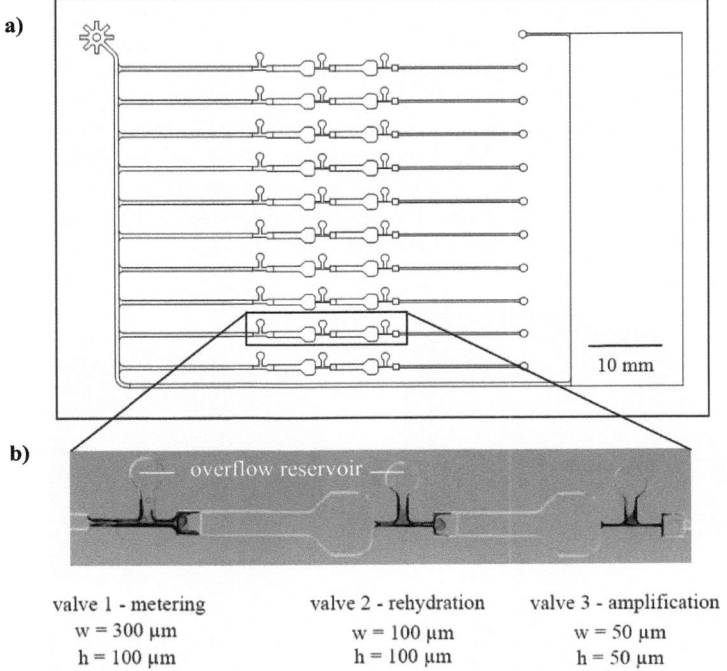

Figure 3.12 (a) Schematic of amplification chip comprising 8 parallel channels with 3 valves each. (b) Close-up of the three high-quality coated valve structures which feature increasing burst pressures due to decreasing dimensions. The sample plug is moved via a static pressure gradient over valve 1 after the sample has been metered and over valve 2 after the reagents have been rehydrated and stays before valve 3 for the amplification reaction.

3.3 Hydrophobic Patterning

Fabrication of Hydrophobic Valves

Due to the small dimensions of the valves, the solution cannot be applied by a standard pipette due to the high risk of overflow. Thus, the valves are coated with a PipeJet-based dispenser [D11]. Here, the dispensing is based on the direct displacement of the spotting solution by squeezing of a polymer tube [127]. With this technology, even particle-containing liquids can be dispensed while exhibiting a low risk of clogging during operation. The applied nozzle features an inner diameter of 200 µm which enables precise dispensing of droplets in the lower nL range. Dispensers and tubes applied for this thesis (please also refer to section 4.1) are depicted in figure 3.13.

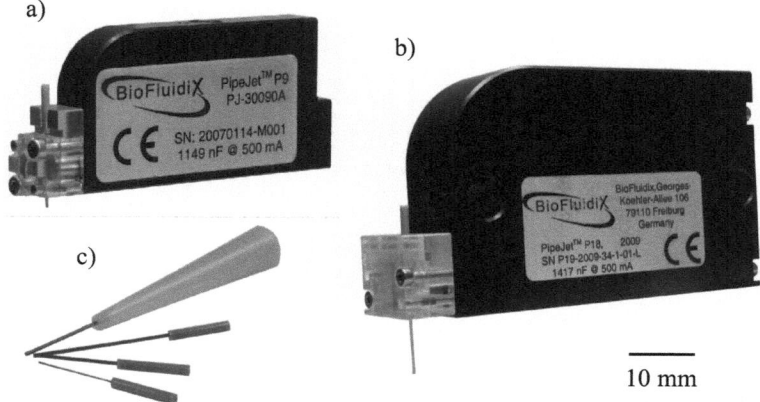

Figure 3.13 Assortment of PipeJet dispensers and tubes. a) PipeJet P9, b) PipeJet P18, c) (top) PipeJet tip, (bottom) different PipeJet tubes, 200 µm - 500 µm inner diameter.

The spotting is optimized by varying the following parameters: single droplet volume, the number of dispensed droplets, the time between dispenses as well as the droplet target position(s). The boundary condition is the minimum coverage required to create a hydrophobic surface in the restriction which translates into a target volume. It would of course be desirable to coat each valve with a single dispense, only. However, as single droplets with volumes > 50 nL can already lead to overflow, it is preferable to dispense a number of smaller droplets instead. Accordingly, the dispenser [D11] is calibrated for the low nL volume range (figure 3.15). Further, if the time between dispenses is set too short (which depends on the vapor pressure of the solvent), the droplets will merge and thus potentially clog the valve. The droplet target position primarily impacts on the risk of overflow when adjacent to the restriction end.

3.3 Hydrophobic Patterning

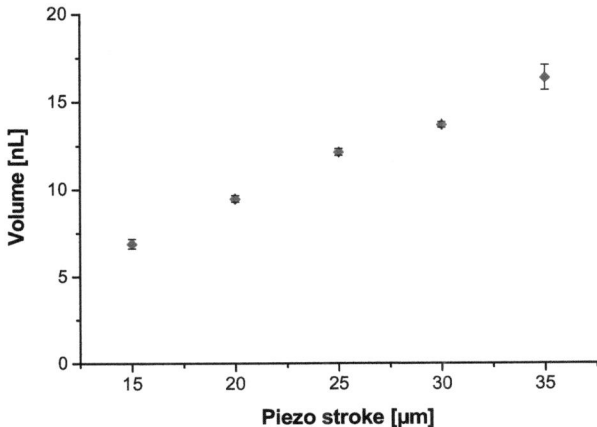

Figure 3.14 Calibration curve for Teflon-CB solution (0.5 wt%) as reference for the hydrophobic patterning featuring a $CV < 5\ \%$ and 7 – 17 nL droplet volumes.

Best patterning results are achieved by dispensing 10 nL droplets of the respective fluoropolymer solution in short succession (~ 1 Hz). To this end, the dispenser is moved by a spotter, i.e. BioSpot 160 [D12], in parallel over the valve structures (figure 3.15) with a velocity of 50 mm/s. Each time the actual coordinate matches one of the pre-programmed valve coordinates, a dispensing is triggered by the software. This procedure is repeated a defined number of times until sufficient coverage of the respective valve is ensured. Then, the dispenser moves to the next row of valves and continues with the patterning. A single chip can be processed in less than one minute.

3.3 Hydrophobic Patterning

Figure 3.15 Amplification chip patterning procedure. The PipeJet P9 is moved in parallel over the valve structures and a dispensing is triggered during movement.

In table 5.3, the estimated surface coverage as well as the coating solutions for valves 1-3 (please refer to figure 3.12) are summarized. Note that more material is applied to the valve surfaces in comparison to the substrates used for the contact angle measurements. This can be explained by the non-uniform valve coating due to wicking of the coating solution after dispensing, i.e. a higher amount is dispensed into the restrictions to ensure superhydrophobicity for the Teflon-CB coating (please refer to table 3.4). Further, for the Teflon-CB coating, solution 5 (table 3.5) is used which requires more dispenses than the highest concentrated solution 6 (table 3.5). However, when using the latter for coating the smallest structures (width $w = 50$ µm), Teflon-CB coated filaments have been observed in the restrictions in some cases. This originates from an increase in solid concentration due to evaporation. To reduce the risk of valve clogging, the lower concentrated solution has thus been used for the hydrophobic patterning.

Table 3.6 Summary of solutions used in experiments (please refer to table 3.5) and the estimated valve surface coverage.

Valve	Solution #	Estimated Surface Coverage
1	1,3,5	150 nL/mm^2
2	1,3,5	150 nL/mm^2
3	1,3,5	200 nL/mm^2

3.3 Hydrophobic Patterning

The required number of droplets as well as spot positions to result in good coverage without risking clogging of the valve structures have been determined empirically for each valve. A step-by-step patterning with a subsequent visual control procedure enabled by the Teflon-CB coating solution has been conducted. The spot positions together with the number of dispensed droplets have been superposed on a close-up of the valve structures featuring a high quality patterning result (figure 3.16).

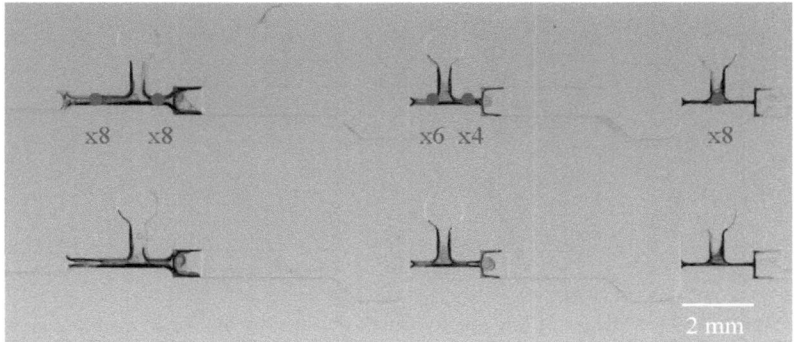

Figure 3.16 Image of patterned valves, spot positions and number of droplets per spot. 10 nL droplets are spotted multiple times on the five spot positions for each row of valves. It should be noted that the exit areas of the valves are intentionally coated to prevent pinning of the liquid meniscus after valve burst.

Quality Control

Multiple experiments have shown that for unoptimized patterning, most of the measurement channels fail to operate properly. Without quality control, each burst pressure would have to be measured to qualify different patterning parameter settings. Thus, the quality control based on dyes enables the rapid optimization of the hydrophobic patterning. The here presented quality control is based on the visual inspection of coated valves. It can either be done *via* microscope or, for the Teflon-CB coating, without any auxiliary material by directly inspecting the chip (for a sufficiently experienced operator). Example images of negative cases for the quality control are depicted in figure 3.17-a, i.e. overflow for valve 1, fluidic shortcut by insufficient coating of the bottom wall for valve 2 and insufficient coverage at the inlet for valve 3. Positive cases are shown in figure 3.17-b whereas the valve structures feature a high quality patterning result. Figure 3.18 depicts an overflow for valve 1 (left) as well as a high quality result (right) for Teflon-QD as coating.

3.3 Hydrophobic Patterning

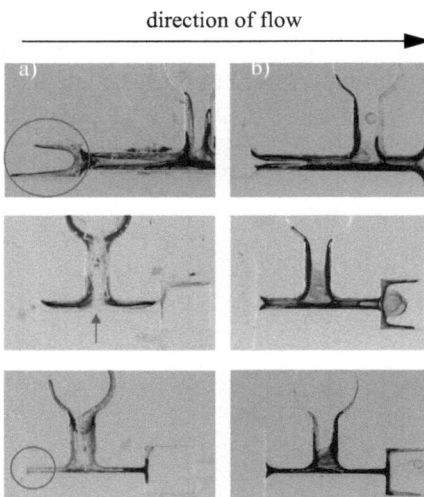

Figure 3.17 (a) Images of low quality valves detected by the visual quality control. For valve 1 (upper left), an overflow into the metering channel at the valve inlet is observable which would lead to a falsely metered sample. For valve 2 (middle left), a fluidic shortcut would occur while for valve 3 (bottom left), the front end of the restriction is not completely covered which would reduce the burst pressure of the valve. (b) Images of high quality valves. Here, the valve structures are completely covered. (c)

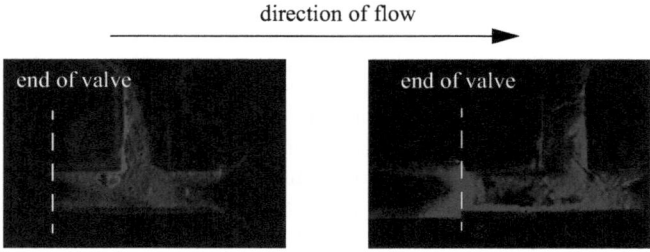

Figure 3.18 Left: image of low quality valve 1 based on Teflon-QD exhibiting overflow. Right: image of high-quality valve 1 based on Teflon-QD coating.

3.3 Hydrophobic Patterning

Burst Pressure Measurements

For determining the burst pressure, each channel is individually connected to the measurement setup (figure 3.19), primed, and sealed. Then, valve B is closed and valve A is opened. The syringe pump is now used to create a defined underpressure. If the hydrophobic valve breaks, a pressure drop is detected by the pressure sensor. The burst pressure corresponds to the pressure level prior to the pressure drop.

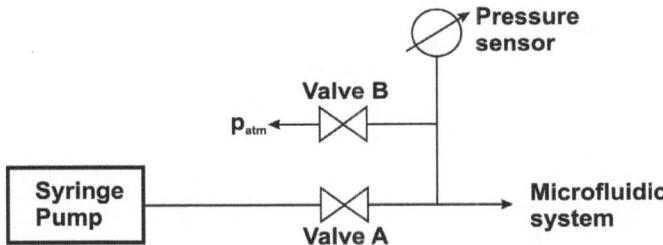

Figure 3.19 Schematic for measuring burst pressures *via* switching valves, pressure sensor and syringe pump.

The measured burst pressures for valves patterned using optimized parameters are summarized in figure 3.20. The parameters have been acquired by visually inspecting Teflon-CB coated valves and varying the single droplet volume, number of droplets, time between dispenses as well as spot position. It can be seen that with the Teflon-CB coating, the highest burst pressures as well as the lowest variability (coefficient of variation $CV < 4\%$ for valve 1, $CV < 7\%$ for valve 2 and $CV < 8\%$ for valve 3, respectively) can be achieved.

The deviation of the theoretical values (please refer to section 2.1.4) could indicate that the radius of curvature of the meniscus is limited for this geometry. The liquid plug is deformed prior to valve burst and the meniscus could get in contact with the restriction walls prematurely when it is pulled inside the restriction. Thus, a coating featuring a contact angle greater than 140° does not further increase the burst pressure of the presented valves.

The higher CVs for the Teflon ($CV < 12\%$ for valve 1, $CV < 13\%$ for valve 2 and $CV < 21\%$ for valve 3, respectively) reflect the inability to do an on-site quality control. Depending on the maturity of the chip processing, it may be required to change the spotting coordinates or number of droplets to compensate for variations in e.g. the hydrophilic coating. This can prove to be quite difficult if no visual feedback is possible.

3.3 Hydrophobic Patterning

For the Teflon-QD, the higher CVs ($CV < 9$ % for valve 1, $CV < 9$ % for valve 2 and $CV < 23$ % for valve 3, respectively) can be explained by the difficult handling of the liquid due to its very fast evaporation. This impacts on the accuracy of dispensing and increases the risk of clogging the dispenser nozzle. A very mature spotting process is therefore required to effectively apply this solution. Also, the burst pressures are lower due to the reduced contact angle. Thus, only in cases where solely a fluorescent dye can be accepted as a method for quality control, Teflon-QD is recommended as coating while Teflon-CB is the clear favorite in all other cases.

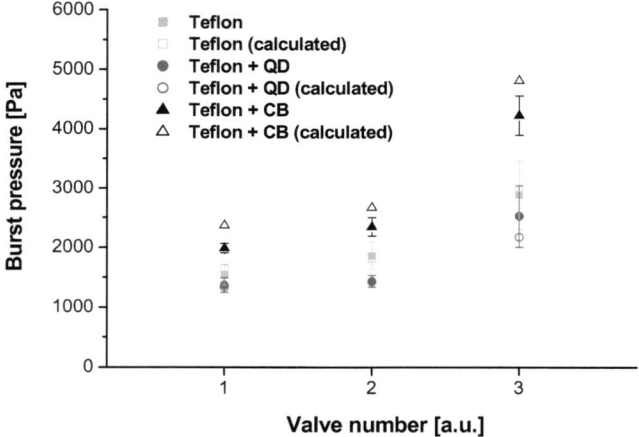

Figure 3.20 Measured and calculated burst pressures of three valves structures and three different coatings. The highest burst pressures and the lowest variability (mean $CV = 6.1$ %) are achieved with the Teflon-CB coating. The higher mean CV of 14.5 % for the solely Teflon coating can be explained by the inability to do an on-site quality control.

Biocompatibility

Teflon-CB as coating has also been tested in respect to biocompatibility by Norchip. To this end, regular PCR strips were coated with Teflon-CB (solution 5) covering small and large surface areas within standard microcentrifuge tubes [M13]. The Teflon-CB solution was applied to the tubes in 1 µl droplets while three different tests were performed. The coated strips were subsequently used to perform NASBA reactions based on the PreTect™ HPV-Proofer kit [77] to evaluate whether the Teflon-carbon black surface would inhibit the enzymatic reaction. A positive control for HPV 16 was used as

3.3 Hydrophobic Patterning

sample material. All tubes containing a reaction mixture (20 µl) with positive control amplified. Thus, it is assumed that for large reaction volumes, the Teflon-CB surface does not inhibit the enzymatic reaction.

Long-term stability

One further interesting aspect is the long-term stability of the coating. To this end, processed and sealed chips which had been stored for 3 months were tested which lead to comparable results for the fluidic functionality in respect to freshly processed chips. This implies that the coating does not degrade or loose its hydrophobicity over time. It should be noted however that the coating is not scratch-resistant, i.e. mechanical abrasion or ultrasonication easily removes the coating. It is thus imperative that the chips are stored accordingly. A tempering step at temperatures $T > 160°C$ could improve the stability of the Teflon-CB coating according to [113]. However, this temperature would exceed the glass transition temperature T_g or even melting temperature of most polymers.

3.4 Conclusion & Outlook

In this chapter, means for the surface modification of polymer labs-on-a-chip have been presented. In detail, the use hydrogen peroxide to effectively sterilize microfluidic chips during pre-treatment is proposed. As hydrophilic coating, it has been evaluated that only PEG exhibits NASBA compatibility and should thus be applied in this scenario. Still, additional coatings have been tested which could be applied for different assays.

With the introduced fluoropolymer-dye solution, namely Teflon-CB, the hydrophobic patterning of microfluidic chips can be significantly improved. First, due to its good visibility, the dried material allows for efficient visual quality control. Second, the coating can create superhydrophobic surfaces on arbitrary polymer substrate materials due to the highly wetting solvent as well as the good adhesion of the coating. If applied on passive valve structures, it can largely increase the respective burst pressures. Third, its biocompatibility should allow for the application in a wide range of lab-on-a-chip-based assays. Compared to the state-of-the-art for hydrophobic valve production (P. Andersson et al. [49]), the presented process does result in less reproducible burst pressures however stronger valves can be produced with a high flexibility. As a consequence, reliable hydrophobic valves can be rapidly produced by simple dispensing without the need of costly clean-room processes which is especially suitable for low-cost labs-on-a-chip based on polymer chips.

To summarize, the presented surface modification steps allow for priming by capillary action and defined flow control while preventing unspecific adsorption. However, especially the hydrophilic coating can cause issues during sealing and the respective compatibility should thus be the focus of future work.

Chapter 4
Reagent Pre-Storage

The pre-storage of reagents for NASBA can be seen as a prominent reference for dry reagent storage on arbitrary labs-on-a-chip as the reaction is very susceptible from a biochemical standpoint due to it's high complexity when compared with conventional PCR [123]. First, the correct ratio of reagent concentrations has to be ensured [123] which requires local deposition of precise reagent amounts. Second, a contamination with RNases has to be prevented in any case [124]. This imposes strict procedures for the handling as well as processing. Another difficulty is the downscaling of the reaction due to an increase in surface effects [100].

For macroscale experiments, commercially produced reagent beads (diameter ~ 5 mm, mass ~ 8 mg) in combination with pipetted primers and probes can be used for the assay. On the amplification chip however, the total sample volume per reaction channel amounts to 640 nL (with a reaction chamber depth of 200 µm) in contrast to 50 µL in macroscale. As reagent beads in these sizes are commercially not available and a distribution of 80 µg powder aliquots into different reaction chambers is near to impossible. Thus, reagent storage on-chip by selective dispensing of reagent solutions into microfluidic chambers is evaluated. Then, reagents can either be dried at room temperature or freeze-dried. Although drying of reagents would be the preferred approach due to its simplicity, experiments conducted at NorChip have demonstrated that dried NASBA-enzymes are not even short-term stable (up to 3 days [125]). As primers and probes do not feature a tertiary structure, simple drying could be feasible for these reagents. Thus, both possible approaches are evaluated in the following sections.

4.1 Principle & Volume Calibration

In preparation for reagent dispensing, stock solutions of the reagents, i.e. mastermix and enzyme mix have to be prepared. For this instance, reagent beads (part of the PreTect-Proofer™ kit) are rehydrated in water-based solutions containing lyoprotectants like BSA according to the protocol for macroscale. The protocol and reagent composition for the mastermix is as follows:

4.1 Principle & Volume Calibration

- Add 80 µl reagent sphere diluent (Tris-HCl, 45% DMSO) to lyophilized reagent sphere (nucleotides, dithiothreitol, $MgCl_2$), vortex immediately to dissolve the sphere
- Add 11,5 of RNase-free water
- Add 5 µl of 10 µM primer 1 (16p1)
- Add 5 µl of 10 µM primer 2 (16p2)
- Add 2,5 µl of 20 µM Molecular beacon probe (16mb)
- Add 16 µl of KCl
- Vortex and spin down

The protocol and reagent composition for the enzyme mix is as follows:

- Add 57 µl of 2 wt% PEG 8000 [M14] solution to the lyophilized enzyme sphere. Let the mixture stand at room temperature for at least 20 min
- Flick tube gently to mix liquid and enzymes

Reagent spotting is conducted with a with a PipeJet-based dispenser [D11] in combination with a disposable tip [D13] as it features a high volume accuracy ($CV < 5$ %). The tip comprises the dispenser nozzle and an injection-molded reservoir (in analogy to a pipette tip) and thus exhibits very low dead volumes. This is especially important due to the high costs of the reagents used (1 mL ~ 1000 €). The spotting is conducted analogous to the hydrophobic patterning procedure (please refer to section 3.3). This reduces the processing time which is highly advantageous for the spotting and freeze-drying (please refer to section 4.3.2). The respective reagent target positions (respective center of reaction chambers) are depicted in figure 4.1 (left: mastermix, right: enzyme mix). For distributing the correct amount of reagents into the respective reaction chambers, a volume calibration of the dispensing process is required. First, the mastermix solution is calibrated (figure 4.2) *via* gravimetrical measurements [D14]. With the applied technology, single droplet volumes of 20-65 nL are accessible. Further, the dispensing features a very good reproducibility ($CV < 2$ %).

4.1 Principle & Volume Calibration

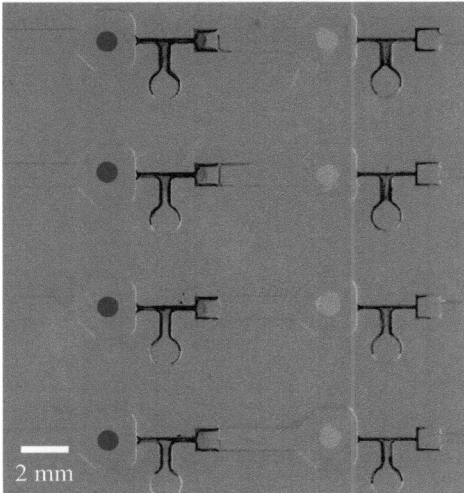

Figure 4.1 Image of patterned valves and adjacent reaction chambers with spot positions for reagents, i.e. left chambers for mastermix and right chambers for enzyme mix.

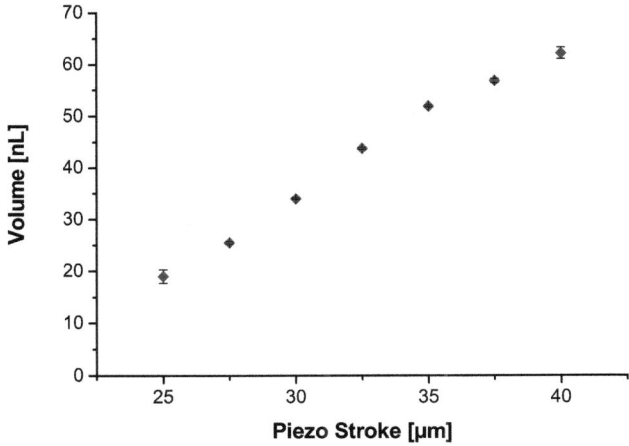

Figure 4.2 Calibration curve for PipeJet-tip based dispensing of mastermix featuring a $CV < 2\ \%$ and a volume range of 20 – 65 nL.

Next, analogous to the mastermix, the dispensing behavior for the enzyme mix is calibrated (figure 4.3) *via* gravimetrical measurements. Here, single droplet volumes from 25-85 nL are accessible ($CV < 4\%$).

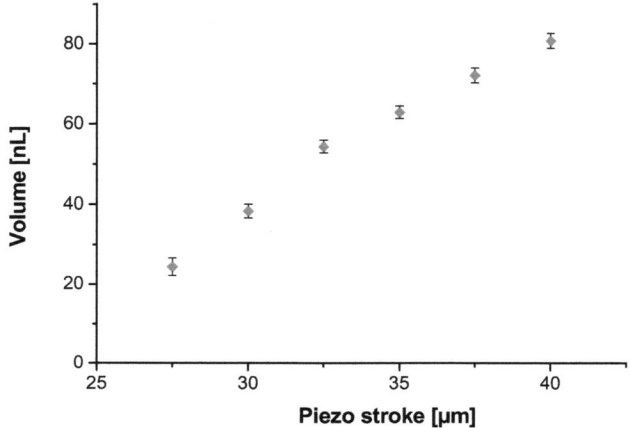

Figure 4.3 Calibration curve for PipeJet-tip based dispensing of enzyme mix featuring a CV of 3.9 % and a volume range of 25 – 80 nL.

For NASBA, it is of great importance not to change the reagent constitution from macroscale as it has been proven to work. Accordingly, the total volume per chamber amounts to 125 nL for the mastermix as well as enzyme mix which thus requires multiple dispenses.

4.2 Reagent Spotting and Drying

Because of the hydrophilic nature of PEG-coated surfaces and the low surface tension of the solutions due to a high surfactant content, it has to be ensured that the dispensed droplet does not come in contact with the walls of the respective reaction chamber after dispensing. This would lead to an undefined spreading of the droplet due to capillary action and consequently an unreproducible reagent deposition. Thus, experiments are conducted to investigate the reagent dispensing and drying with the use of fluorescent dyes (60 µg/mL FITC-labeled antibodies and 2 mM FAM-labeled DNA oligos for the enzyme- and mastermix, respectively, provided by Norchip) for better visualization. As experimental setup, an inverted microscope [D1] with a custom-made excitation source [D15,D16] and CCD camera [D2] is used.

4.2 Reagent Spotting and Drying

4.2.1 Spotting into Reaction Chambers

Enzyme Mix

The enzyme spotting is evaluated by acquiring fluorescence images of a reaction chamber after dispensing at selected time intervals (1, 2, 3, 5 and 60 s) for a fixed droplet volume (~ 25nL). The position of the dispenser nozzle is adjusted manually on the chamber center by referencing the position to the fluorescing test liquid in the nozzle. As can be seen in figure 4.4, the droplet stays well inside the reaction chamber for the selected droplet volume of 25 nL. The test sample spreads in the reaction chamber for about 5 seconds after which a steady state is reached.

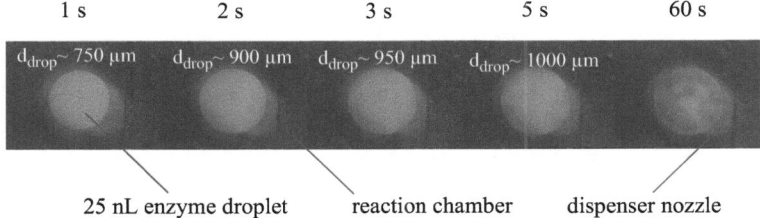

Figure 4.4 Example image series for dispensing enzyme mix droplets (25 nL) into the reaction chamber after selected time intervals (same intervals applicable to subsequent image series). The brightness has been increased for better visibility of the chamber boundaries. Apparently, the droplet spreads after contact with the chamber surface but stays well off the walls.

Next, the procedure is repeated for five different reaction chambers (not shown). Again, the droplets stay well inside the reaction chamber but a variable spreading of the test liquid (+/- 100 μm) can be observed which can be accounted to the inhomogeneity of the PEG-coating. An important question which has to be considered is how a deviation of the spot position from the center of the reaction chamber can impact on the reagent spotting and drying. Different effects have to be taken into account:

- Dispenser nozzle position offset
- Deviation of the droplet flight path due to dried sample at the nozzle orifice [126]
- Deviation of the droplet flight path due to fast movement of the dispenser in combination with target distance
- Random deviation of the droplet flight path
- Positioning accuracy of the spotter

4.2 Reagent Spotting and Drying

The first effect can be counteracted by spotting on a reference position and adjusting the target coordinates based on this offset. The second effect can be addressed by a tight control of the processing, i.e. by preventing downtimes. The third effect is controllable with optimized process parameters. The droplet velocity after ejection typically amounts to about 1 m/s [127]. With a target distance of about 5 mm, the axis velocity of the spotter should not exceed 20 mm/s to keep the deviation at or below 100 µm. Triggering the dispensing prior to reaching the center of the reaction chamber would compensate for this effect and allow for faster processing. However, the risk of the droplet hitting the chamber walls would increase. Taking into account the random deviation which is estimated to be +/- 50 µm and the positioning accuracy of the spotter (better than +/- 100 µm), the superposed accuracy amounts to - 150 µm / + 250 µm and should thus be adequate for accurately spotting the enzyme mix into the respective reaction chambers. As control, a misalignment of at least 500 µm was required to result in undefined wetting along the chamber wall (figure 4.5).

Figure 4.5 Image series for dispensing enzyme mix droplets (25 nL) into a reaction chamber with a manual misalignment of ~ 500 µm. Immediately after the droplet gets in contact with the chamber boundary, it gets pulled along the walls by capillary action.

As the current target volume for the enzyme mix spotting amounts to 125 nL, five droplets of 25 nL have to be dispensed subsequently with sufficient time for each droplet to dry. As a consequence, the drop-in-drop dispensing (2nd and 3rd dispense) has been evaluated with a time delay of ~ 60 s between dispenses (figure 4.6). As can be seen from the image sequences, the subsequently dispensed droplets are retained inside the form of the first droplet as the dried reagents act as a fluidic barrier. This also applies for the 4th and 5th dispense (not shown).

4.2 Reagent Spotting and Drying

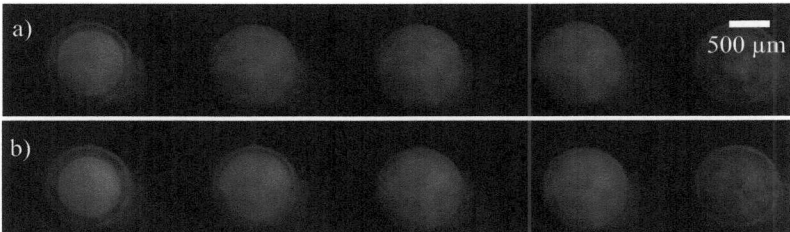

Figure 4.6 Image series for dispensing enzyme mix droplets (25 nL) into dried droplets (2nd and 3rd dispense, a) and b), respectively). Here, the dried reagents act as barrier for the subsequent droplets while staying inside the droplet boundaries.

Next, it is evaluated if larger droplet volumes are feasible as it would allow for faster processing. Thus, a volume of 50 nL and 75 nL is dispensed into a chamber (not shown). For 50 nL, the droplets still stay off the wall although the margin for misalignment is effectively reduced. For a volume of 75 nL, the droplets come in contact with the chamber wall during spreading. It is thus proposed to dispense 25 nL droplets into the reaction chambers.

Mastermix

Again, a fixed dispensing volume of 25 nL is selected and the mastermix spotting is evaluated by acquiring fluorescence images (figure 4.7, 1st and 2nd dispense) of the reaction chamber after dispensing at selected (i.e., similar) time intervals. The results for the mastermix solution are basically comparable with the ones obtained for the enzyme mix. Again, the volume should be kept at 25 nL to allow for error compensation.

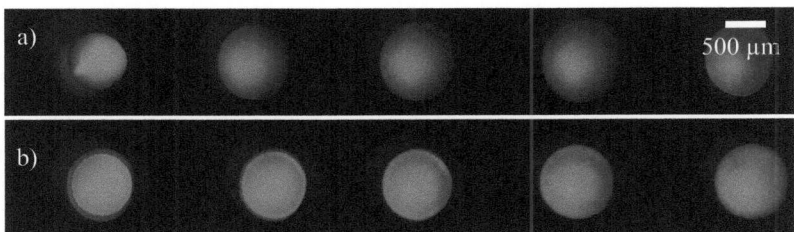

Figure 4.7 (a) Image series for dispensing mastermix droplets (25 nL) into a reaction chamber. Again, the droplet spreads after contact with the chamber surface but stays well off the walls. (b) Image sequence of the 2nd dispense into a dried droplet. Again, the droplet stays in the boundary of the first dispense.

4.2.2 NASBA Results

As proof-of-principle for spotting and drying, a mixture of primers and probes (please refer to section 4.1) as sample liquid for the mastermix is spotted into respective reaction chambers, 125 nL each, on at least ten different chips. Then, the chips are sealed *via* adhesive foil [M12] and sent to Norchip for conducting NASBA on-chip. Figure 4.8-a depicts an example image of reaction chambers with spotted primers and probes which have been used for the experiments. Primers and probes specific to HPV 16, HPV 31 and HPV 33 were spotted into different chips to demonstrate the capability for parallel NASBA on-chip. NASBA results are considered positive if the measured fluorescence signal exhibits the characteristics of a real-time amplified reaction [123]. As can be seen in figure 4.8-b for e.g. a positive clinical sample of HPV 33, all chambers feature amplification. Comparable results could also be achieved for HPV 16 and HPV 31 (not shown). Further, the results have been be reproduced after 2.5 months of chip storage.

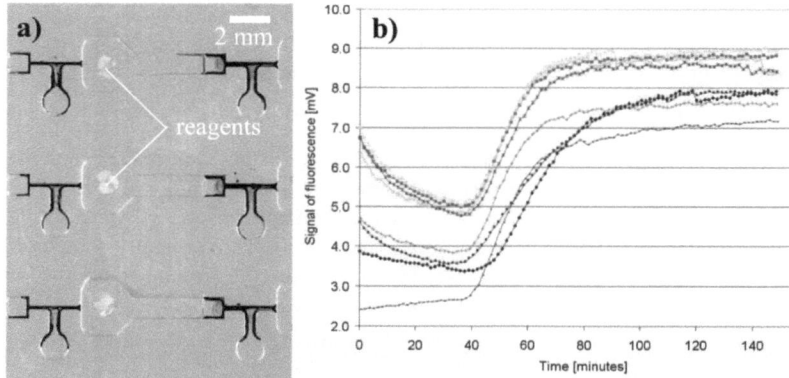

Figure 4.8 (a) Image of typical reagent cakes, i.e. spotted primers and probes, in reaction chambers. (b) NASBA characteristics for 8 channels (single chip, HPV 33) all exhibiting positive amplification.

4.3 Reagent Spotting & Freeze-Drying

Before freeze-drying, the reagents first have to be frozen. Due to the small total reagent volumes used for the amplification chip (125 nL), freezing of liquid reagents on-chip after dispensing is not practical as most of the liquid has already evaporated at this time (t_{eva} < 1 min [128]). Also, this should be seen as a general issue for sub-µL volumes of reagent solutions. It is however possible to cool down microfluidic chips prior to the spotting which results in a quasi-instantaneous freezing of the reagent droplets on the chip surface upon impact.

4.3.1 Spotting on Frozen Substrate

The amplification chip as well as an aluminum chip holder is put into a -80°C freezer for approximately 2 hours. Subsequently, the cooled holder including the chip is refastened on the spotter and reagents are dispensed into the reaction chambers (please refer to section 4.1). Storing the chips in petri dishes prevents the development of ice crystals on the surface during freezing which would increase the risk of RNase contamination. Before and after spotting, the chips are stored on dry ice to keep them frozen. If the chip spotting is not conducted rapidly (> 60 s), thawing will occur. This could be counteracted by adding an active cooling unit to the spotter. However, the high temperature difference between chip and ambient air would still lead condensation. Only working in an atmosphere with an ambient temperature below zero would prevent this effect.

As preliminary test, 25 nL droplets of enzyme mix are dispensed into the reaction chambers of a chip, first a single droplet per chamber (figure 4.9), then four droplets subsequently (not shown). Upon impact, the droplet freezes and is therefore unable to spread or touch the walls if ruling out strong misalignment (figure 4.9). For dispensing of multiple droplets, the subsequent droplets only come in contact with the previous droplet(s) and tiny pillars are formed. The reagent pillar can however exceed the depth of the chamber which poses a problem for subsequent sealing.

4.3 Reagent Spotting & Freeze-Drying

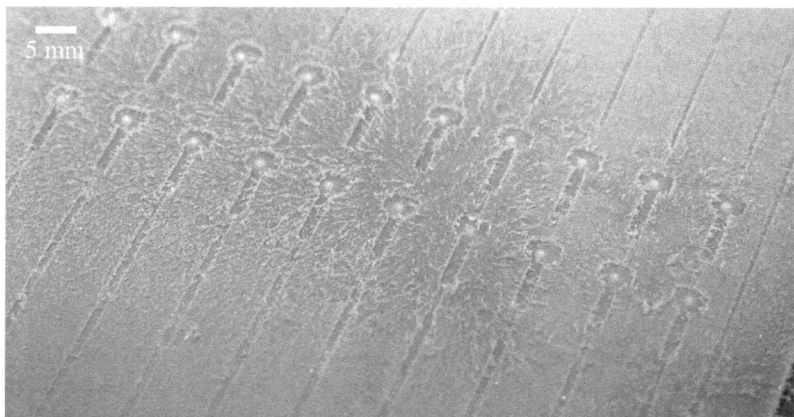

Figure 4.9 Example image for dispensing 25 nL droplets of enzyme mix into reaction chambers of a frozen test chip. It can be observed that all droplets stay well in the center of the respective chambers. Condensed water from the atmosphere also freezes when getting in contact with the chip surface.

In order to prevent these pillars from touching the sealing foil, spotting into each reaction chamber is conducted at two different positions (not shown). Further, the single droplet volume can be increased to ~ 60 nL due to the non-existent spreading of droplets.

To summarize, this method has not only the advantage of possibly increasing the long-term stability of stored reagents, it also increases the allowed margin of error of the reagent spotting. One downside of this method is the very short processing window which requires an experienced operator. For mid-scale chip production, an active cooling solution would be imperative. Further, water condenses on the chip surface due to the high temperature difference between chip surface and ambient air. As this water is not necessarily RNase-free, it could possibly negatively impact on the test results. A countermeasure however would be the use of a clean bench for spotting.

4.3.2 Freeze-Drying

Basic Theory

The freeze-drying process which is used for dehydration of a sensitive material is based on the sublimation of frozen liquid in a sample, i.e. the direct transition from solid phase to gas [51]. At first, the solution has to be frozen whereas the freezing time defines the ice crystal size as well as the product morphology after drying. Again, it should be pointed out that due to very small sample volumes used for the amplification chips and thus a very high surface to volume ratio, it is not possible to conduct a slow freezing of the sample. Then, the frozen reagents are dried *via* sublimation, first by primary drying, then by secondary drying [51]. During primary drying, the freeze-drying cavity is evacuated and enough heat is supplied either by heat radiation or by heated plates for the liquid to sublimate. In this initial drying phase, about 98% of the liquid in the material is sublimated. Furthermore, a typical freeze-dryer features a cold condenser which provides a surface for the liquid vapor to re-solidify on. The secondary drying phase aims to sublimate liquid molecules that are bound by physical-chemical interactions. In this phase, the temperature is raised in respect to the primary drying phase and a stronger vacuum is applied in to enforce sublimation. After the freeze-drying process is complete, the vacuum is usually broken with an inert gas such as nitrogen.

Instrumentation

For freeze-drying of reagents on the amplification chip, a standard freeze-drier is applied [D17]. It features an ice condenser temperature of -88°C, the capacity to freeze-dry up to 4 kg of material and a program module for the automatic execution of custom process protocols.

Experimental Protocol

The experimental procedure has been determined together with Norchip by conducting a test run with only enzymes to reduce the complexity. Currently, it is based on the storage of freeze-dried reagents at -18°C and not at ambient temperature as a better long-term stability is expected [129].

- Store patterned and coated test chips in -80°C freezer
- Use petri dishes to protect from surface contamination
- Freeze chips for 2 h
- Transport chips in dry ice from freezer to spotter
- Spot chips (2 droplets of enzyme solution,~ 63 nL per droplet)
- Transport chips in dry ice from spotter to freezer

4.3 Reagent Spotting & Freeze-Drying

- Cool down freeze-dryer to -88°C
- Transport chips in dry ice from freezer to freeze-dryer, put on heatable plate
- Start freeze-drying procedure (please refer to the next paragraph)
- Stop freeze-drying procedure, cooler stays on
- Introduce nitrogen (slowly!) until freeze-dryer at atmospheric pressure, keep cavity under constant flow of nitrogen afterwards
- Seal chips with adhesive foil in freeze-dryer cavity
- Put chips in protective bag, store chips in dry ice for transport

It should be pointed out that sealing the chips in the freeze-dryer cavity greatly reduces the risk of reagent rehydration as the high temperature difference between the freeze-dryer and ambient atmosphere promotes condensation if removing the chips without a protective cover. Even when applying a sophisticated sealing solution after freeze-drying, the chips should be temporarily sealed or put in protective containers to prevent this effect.

Freeze-Drying Procedure

The basic freeze-drying procedure has been derived from a patent [130] as similar reagent compositions were freeze-dried. Because the total reagent volume dispensed into the amplification chips is very small (~ 10 µL in contrast to ~ 1 mL), much shorter drying times are required although a longer time is required for fast frozen materials in contrast to slowly frozen materials [131]. Thus, based on an empirical assumption, the time at constant temperature is reduced by a factor of 8 and the times where the temperature is ramped up is reduced by a factor of 4 as these steps have to be considered critical. The vacuum pressures are also adapted based on a discussion with a technical expert from the freeze-dryer manufacturer. The proposed freeze-drying procedure is summarized in table 4.1

4.3 Reagent Spotting & Freeze-Drying

Table 4.1 Freeze-drying protocol, split into supply, primary drying and secondary drying. First, the pressure is lowered for constant temperature at the start of the primary as well as secondary drying. Then, the temperature is ramped up and held for some time. Chips are introduced after step 2.

Step #	Type	Pressure (mbar)	Temperature (°C)	Time (min)
1	Supply Materials	1000	-35	-
2	Freezing & Pump warm-up	1000	-35	30
3	Primary Drying	0.10	-35	20
4	Primary Drying	0.10	-10	30
5	Primary Drying	0.10	-10	45
6	Primary Drying	0.020	-10	10
7	Secondary Drying	0.020	10	15
8	Secondary Drying	0.020	10	30
9	Secondary Drying	0.020	25	120

A test chip with partially freeze-dried enzymes (double amount, 250 nL, i.e. four 63 nL droplets of enzyme mix per reaction chamber) is depicted in figure 4.10. The image has been acquired during freeze-drying in the freeze-dryer cavity. After removing the test chips from the freeze-dryer cavity, the morphology of the reagent pillars stays the same and no rehydration could be observed if sealing the chips with a protective foil. If the height of the enzyme pillars exceeds the depth of the chamber, the reagents get in contact with the sealing foil where they tend to stick (figure 4.11). In itself, this in not a major problem as the foil material is basically biocompatible but if targeting a sophisticated sealing solution, the foil has to be removed (if not using a protective container) thus also removing part of the reagents. Additionally, for temperature diffusion bonding as sealing solution, the enzymes will experience an increased temperature impact (please refer to section 5.1.3).Consequently, this issue will have to be addressed by e.g. dispensing smaller volumes on different positions or by increasing the reaction chamber depth.

4.3 Reagent Spotting & Freeze-Drying

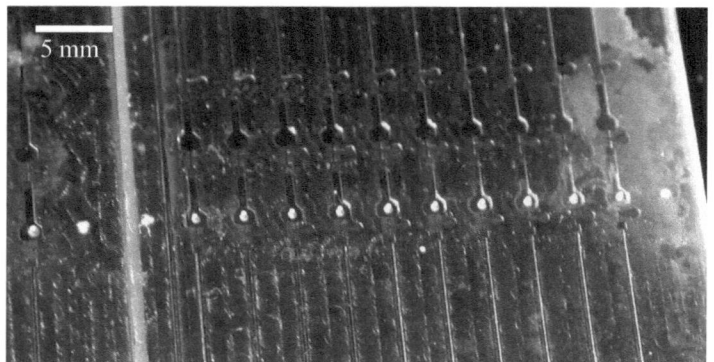

Figure 4.10 Image of amplification chip in freeze-drier cavity featuring partially freeze-dried enzymes in the reaction chambers.

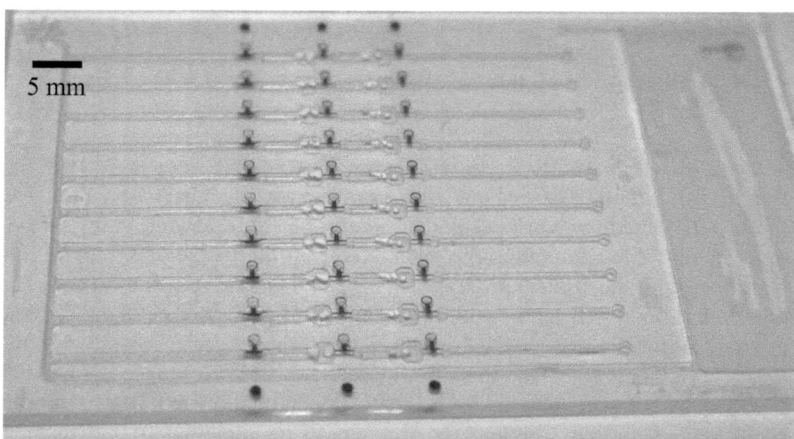

Figure 4.11 Image of sealed amplification chip featuring freeze-dried mastermix and enzymes in the respective reaction chambers. Here, the reagents get in contact with the sealing foil which is not desirable.

4.3.3 NASBA Results

As proof-of-principle for the spotting and freeze-drying, 125 nL of enzyme mix is spotted into respective reaction chambers on different chips. Then, the chips are processed in accordance with the experimental protocol and sent to Norchip for conducting the NASBA reaction. As mentioned in the preface of this chapter, air dried enzymes did only feature a long-term stability of 3 days. For the freeze-dried reagents, successful reactivation and amplification could be demonstrated (figure 4.12) after 3 weeks. As these experiments were conducted before introducing the hydrogen peroxide treatment step to effectively remove RNases (please refer to section 3.1.2), a longer shelf life should be achievable based on the adapted protocol.

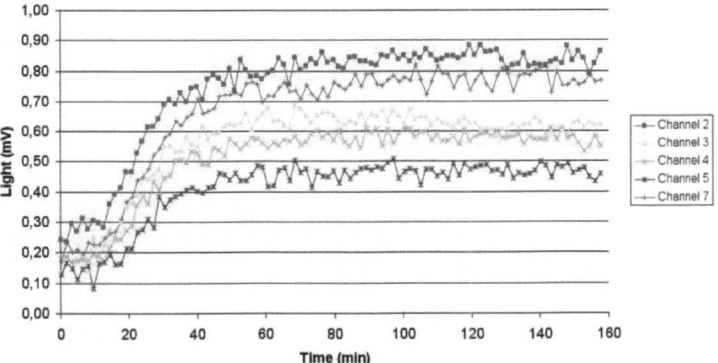

Figure 4.12 NASBA characteristics for HPV 16 based on freeze-dried enzymes exhibiting positive amplification after 3 weeks of storage.

4.4 Reagent Bead Fabrication

The presented method of spotting and freeze-drying has certain disadvantages, e.g. the need for chip freezing, the limited number of chips processable in parallel and the danger of rehydration between freeze-drying and sealing. Lyophilized beads [132] that could be placed into the reaction chambers before the sealing would address these issues. As a consequence, it is evaluated if freeze-dried reagent beads with diameters down to 150 µm can be fabricated and distributed into the reaction chambers.

A common approach for reagent bead production is the dispensing of reagent solutions into liquid nitrogen [133]. The reagents freeze in the carrier liquid after contact and spherical particles are formed due to surface tension effects. An alternate approach is the use of a perfluorinated solvent (e.g., Fluorinert

4.4 Reagent Bead Fabrication

[M7]) cooled down to -80°C [134]. As the solvent features a specific density of ~ 1.8 g/cm^3, the reagent beads float on the solvent surface and can be easily removed. Accordingly, reagent droplets are dispensed [D11] into a liquid nitrogen bath. Depending on the dispensing volume and the flight path (i.e., evaporation time), beads with a diameter 150-400 μm diameter have been produced (e.g., Figure 4.13). The liquid nitrogen is sublimated during lyophilization thus leaving the freeze-dried beads on the bottom of the carrier. Unfortunately, electrostatic charges complicate the handling resulting in the inability to transfer the beads into the respective reaction chambers. Electrostatic charges are generated when trying to move the reagent beads from the carrier. This can be explained by the triboelectric effect [135] which leads to an electron transfer between adjacent materials. Furthermore, especially the mastermix beads feature a high surfaces porosity, resulting in an increased risk of rehydration. Concluding from the experiments, beads of the required size can be produced. However, further process development to improve the surface morphology or change the reagent constitution in respect to the triboelectric series [135] (i.e., include salts or sugars to adapt the triboelectric amplitude or polarity to that of the carrier) has been stalled due to time constraints.

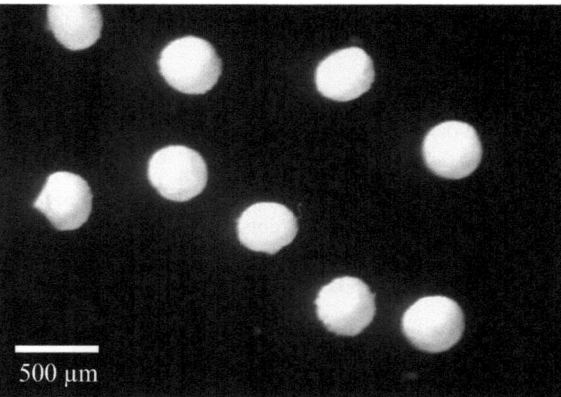

Figure 4.13 Image of freeze-dried enzyme beads fabricated by dispensing enzyme mix into liquid nitrogen.

4.5 Conclusions & Outlook

In this chapter, two approaches for the pre-storage of bioreagents have been presented and evaluated, namely spotting and drying as well as spotting and freeze-drying.

With a droplet volume of 25 nL, both reagent mixes are successfully dispensed into PEG-coated reaction chambers while preventing any contact with the adjacent walls for the spotting and drying. It is also demonstrated that by dispensing multiple droplets in series into the same spot with an interim drying step, the target volume of 125 nL for the respective solutions can be safely deposited into the reaction chambers. The concept is verified by spotting and drying primers and probes specific to three different HPV types. With the spotted chips, a successful NASBA amplification based on clinical samples has been conducted even after 2.5 months.

For spotting and freeze-drying, the freezing of the reagents has been realized by spotting on frozen chips. When dispensing multiple droplets onto the same spot, the subsequent droplets come in contact with previously frozen droplets, only, consequently forming small pillars. These pillars have been subsequently freeze-dried on-chip using a custom-developed protocol. Reactivation of susceptible enzymes has been demonstrated after three weeks. In contrast to previous work by M. Brivio et al. [52], the reagent volume has been reduced by more than one order of magnitude. Further, the presented concept allows for flexible positioning of the reagent spots.

To summarize, both methods are applicable for reagent storage however only freeze-drying ensures sufficient enzyme activity after mid-term storage, i.e. three weeks. Consequently, future works has to focus on extending the shelf life of freeze-dried enzymes to permit long-term storage, i.e. months.

Chapter 5
Biocompatible Sealing

Independent of the means of fluid transport, i.e. capillary action [30], pumping [25] or centrifugation [26] for continuos microfluidics or electrowetting [136] for droplet-based microfluidics, microfluidic chips have to be sealed to be able to operate. The selection of appropriate sealing technologies for labs-on-a-chip mainly depends on the biocompatibility of the required materials and processing techniques. Biocompatibility describes the quality of a material of not having any toxic or hazardous effects on biological systems [137]. Due to the high surface-to-volume ratio in microfluidic chips, undesired effects related to material biocompatibility occur more frequently [138]. The selected chip material (COC) is regarded biocompatible as bulk material [139]. However, the channel surface can impact on the functionality of reagents since proteins are easily adsorbed. Thus, the sealing process must not only feature biocompatible bulk materials and process conditions but also work with coated channels (please refer to section 3.2). Additionally, it must not destroy any pre-stored reagents (please refer to chapter 4).

Consequently, two sealing methods which are inherently not incompatible with the described boundary conditions (please refer to section 1.2.3), namely the temperature diffusion bonding and the adhesive bonding have been selected. In the following sections, the principle of these bonding methods will be discussed. Then, results from the process development and evaluation will be presented, concluding with an optimized process chain for both processes.

5.1 Temperature Diffusion Bonding

5.1.1 Principle

Typically, the process of temperature diffusion bonding uses two identical or similar polymer materials featuring the same T_g as bonding partners. A tight seal is established by the heat induced crosslinking of polymer chains at the interface between substrate and sealing foil. The diffusion of polymers can also be explained by Ficks' law since a long contact time and the induction of high temperatures allows for diffusion of polymer chains [140]. The disadvantage of this process is the thermal deformation and potential sagging of the lid which can possibly lead to channel obstruction, especially when considering low aspect ratio geometries. As a consequence, the base process has been adapted in a previous work [59] and a so-called compound foil featuring a thick, mechanically stable cover foil with a high T_g and an adhesive

5.1 Temperature Diffusion Bonding

layer featuring a low T_g is used (figure 5.1). Here, only the adhesive layer is melted during lamination thus preventing thermal deformation of the cover foil. In this work however, the applied cover foil is not transparent which prevents read-out based on fluorescence. Thus, the process is again adapted to feature a transparent foil.

In respect to equipment, a laminator featuring a top heatable roll as well as the bonding partners are required. The bond quality is influenced by the roll pressure p_{roll}, the roll temperature T_{roll}, and the velocity of the rolls v_{roll} which move the substrate through the laminator. A protective foil should be placed on top of the bonding partners to prevent damaging of the upper roll.

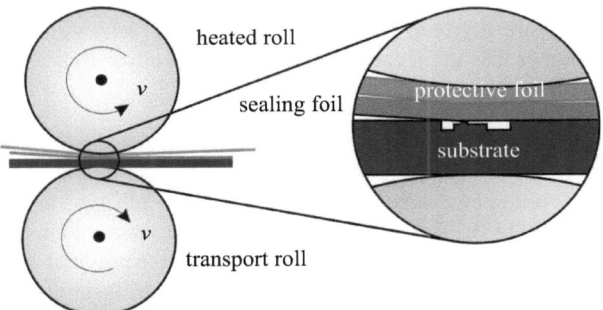

Figure 5.1 Principle of temperature diffusion bonding. Both substrate and sealing foil are moved with constant velocity v and brought into intimate contact with a pressurized roll system where the top roll is heated. A protective foil is placed on top of the bonding partners.

5.1.2 Materials and Methods

Foil Material Selection

Besides biocompatibility, requirements for the sealing foil or lid are superior optical properties like high light transmission and low autofluorescence combined with a good mechanical stability. The former is imperative to enable optical read-out techniques and can be addressed by the exclusive use of amorphous olefin polymers [139,141]. For achieving the latter, foils exhibiting a thickness > 150 μm should be applied. To the author's knowledge, no high-T_g COC (e.g. Topas [M1]) foil is available on the market with a thickness exceeding 100 μm. Thus, a cyclic olefin polymer (COP) foil [M15] which is available as off-shelf product is selected as cover foil material. The material properties are summarized as follows:

- Glass Transition Temperature: 136° C
- Light Transmission: 92 %
- Tensile Strength: 60 MPa
- Foil Thickness: 188 µm

To guarantee the survival of pre-stored reagents on-chip, high temperature impact during the sealing process has to be avoided. Therefore, an adhesive layer with a low T_g is desirable for the process. The most feasible solution is the use of another olefin polymer variant as adhesive layer because it exhibits roughly the same optical properties as the foil and substrate material. Thus, COC [M16] with a T_g of 75°C has been selected. The resulting foil comprising the two polymer materials ensures a temperature diffusion of the polymers at an induced temperature between the polymer layers of 75°C without any thermal deformation or loss in mechanical stability of the cover foil.

Foil Production

For producing the compound foil, spin-coating is selected to apply the adhesive layer onto the cover foil. Spin-coating is a simple method to apply uniform solid films onto plane substrates [142]. Thus, the adhesive layer has to be in liquid form [143]. The solution of polymers is controlled by the dissolution of polymer chains or the diffusion of chains at the polymer-solvent interface. All amorphous olefin polymers are resistant to hydrolysis, acids, and most of the common polar solvents like alcohols but exhibit solubility in chlorinated or non-polar, aromatic compounds such as toluene and naphta. However, an insufficient solvation of polymer macromolecules leads to deviations and inhomogeneities when applied as a thin layer [144].

In this work, toluene has been used as solvent. Due to the high volatility and biodegradability of toluene, it has a short half-life on the foil surface and is biocompatible [145]. The solution should be prepared in a closed bottle under constant stirring until the polymer is completely dissolved ($t > 2$ h). As illustrated in figure 5.2, the COC solution is dispensed onto the rotating substrate in order to disperse the liquid by centrifugal force. The centrifugal force causes most of the solution to spread to and over the edges of the substrate. The final film thickness primarily depends on the rotational speed and fluid viscosity [146]. The best spin-coating results are obtained by using a 5 wt% solution of COC [147] which is thus used for the compound foil fabrication. For rectangular chips, the spin-coating of a CD-shaped foil is not feasible. The excision of the foil leads to mechanical stress on the material which results in micro cracks in the surface of the compound foil. A pre-cut foil (figure 5.2) prior to the coating avoids these negative effects and leads to similar adhesive layer thickness with the exception of the foil edges [148].

5.1 Temperature Diffusion Bonding

The spin-coating parameters for a CD-shaped foil [59] are used as basis for the process development. Thus, the rotational velocity and total spinning time of the spin coater is varied and the resulting layer thicknesses is measured with a profiler [D18] subsequently. Then, the bond quality in respect to the layer thickness is assessed. Best results are achieved with the deposition of 5 mL polymer solution and a rotational velocity of 1000 rpm for 20 seconds resulting in a layer thickness of 6 µm +/- 0.5 µm. Foils coated with higher rotational velocity (layer thickness < 4 µm) exhibit weaker or almost no bonding.

Figure 5.2 Spin-coating process of rectangular foil comprising: (a) applying of solution (b) spinning of sample (c) drying and evaporation.

Device

As device, a custom-built laminator [D19,149] is applied (figure 5.3). The temperature of the two heatable rolls can be set up to 200°C while the roll velocity can be set from 0.3 to 3 m/min. Additionally, the roll pressure and distance to target can be easily adjusted.

Figure 5.3 Custom-built laminator. The device allows for reproducible temperature, pressure, and roll velocity control.

Lamination Process Development and Protocol

To avoid any damages of the upper heatable steel roll, a 100 μm thick aluminum foil is placed on top of the bonding partners. Additionally, putting a 100 μm thick clean-room cloth between the protective layer and bonding partners prevents a deformation of chip and lid due to extensive pressure respectively temperature impact.

The next step in temperature diffusion bonding process development is the evaluation of optimized process parameters. The temperature of the upper roll is not equivalent to the temperature at the interface of the bonding partners since the chip is moving with a constant velocity and protective layers slow down the heat transfer. As the T_g of the adhesive amounts to 75°C, it can be assumed that the interface has to reach at least this temperature to form a stable bond. Accordingly, by varying the temperature and the velocity, starting with $T_{min} = 125°C$ and $v_{min} = 0.3$ m/min, bonding parameters which provide a stable bond and minimal temperature impact are determined. The applied pressure during the process should not exceed 2 bar to prevent stress-cracks on the foil. The roll distance depends on the thickness of the chip holder used in the process and should be set 0.5 mm less than the total height of the holder/chip assembly. Further, a second laminating cycle is necessary to achieve a stable bond. Parameters which result in very good bonding results (referred to as optimal parameters) are $v_{roll} = 0.4$ m/min, $T_{roll} = 140°C$ and $p_{roll} = 2$ bar.

5.1.3 Evaluation

Temperature Test

For temperature diffusion bonding, denaturation of bioreagents by too high temperatures is a major issue which has to be taken into consideration. Even if the T_g of the adhesive amounts to 75°C, stored reagents are typically located on the bottom of channels and ambient air provides very good isolation due to it's low thermal conductivity. For NASBA reagents, experiments conducted by Norchip have demonstrated that dried enzymes are allowed to experience temperatures up to 65°C for a time span of 2 minutes. These results should be representative for arbitrary pre-stored enzymes.

Two different tests for characterizing the temperature impact have been selected and will be described in the following section. To determine the temperature *via* temperature sensor, a thermocouple [D20] is glued into structures of different depths (figure 5.4). It features fast response times (< 300 ms), easy assembly and small dimensions. The temperature during lamination is monitored by a digital panel meter [D21].

5.1 Temperature Diffusion Bonding

Figure 5.4 Experimental setup for the temperature test *via* thermocouple. The temperature sensor is sequentially glued into channels of different depths and the temperature is monitored during lamination multiple times for each depth.

The optimal parameters are applied and the bonding partners together with the temperature sensor are moved through the laminator. The experimental results are displayed in figure 5.5. The mean temperature at the interface amounts to 80°C which is above the T_g of the adhesive compound foil and thus meets the temperature requirement for bonding. At the bottom of a 200 μm deep cavity, the mean temperature is 61.1°C ± 4.1°C which is only marginally higher than the acceptable temperature of 65°C in the worst case. The time for laminating the amplification chip amounts to 15 seconds per run.

5.1 Temperature Diffusion Bonding

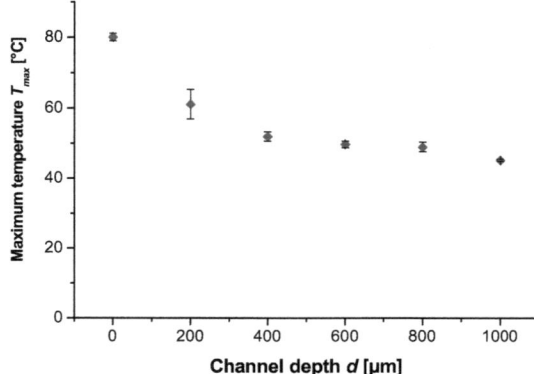

Figure 5.5 Experimental results of the thermocouple-based temperature tests. At the interface, the temperature reaches 80° C, higher than the required T_g of 75°C. At the specified channel depth, namely 200 µm, the temperature stays at 61° C, i.e. below the critical temperature of 65° C.

Another method for determining the temperature impact on the reagents are so-called temperature indicators. Generally, they are designed for measuring the surface temperature by marking the surface to be heated. This effect corresponds to the melting point of the indicator and is therefore irreversible. The indicator is initially opaque but liquefies and becomes transparent as soon as the rated temperature is reached.

The selected indicator [D22] features a melting temperature of 65°C (temperature tolerance ± 1%). The indicator simulates dried bioreagents with an approximate height of 20 µm deposited in 200 µm deep channel. For the precise deposition of 20 µm thick indicator layers inside a channel, the indicator powder is dissolved in toluene. With a determined mix ratio of 1:10 (10 ml toluene, 100 mg indicator) the dispensing of 2 µl solution results into a layer thickness of ~ 20 µm after the solvent is evaporated. In a first step, the solution is dispensed into a channel of 200 µm depth and an image of the dried indicator is captured [D10,D2]. Then, the test chip is laminated based on optimal parameters. Another image is captured and compared to the one before lamination. Thus, qualitative information if the dried reagents have experienced too high a temperature can be derived from the resulting images.

Three representative images of the indicator material on-chip are depicted in figure 5.6. Figure 5.6-a displays the indicator before lamination while figure 5.6-b shows the indicator after lamination whereas only slight deformations are observable. For positive control, the chip has been manually heated up to 65°C and a clear change in morphology can be observed (figure 5.6-c). With the above results, it can be assumed that the indicator has not experienced more than the characteristic temperature of 65°C during the sealing process.

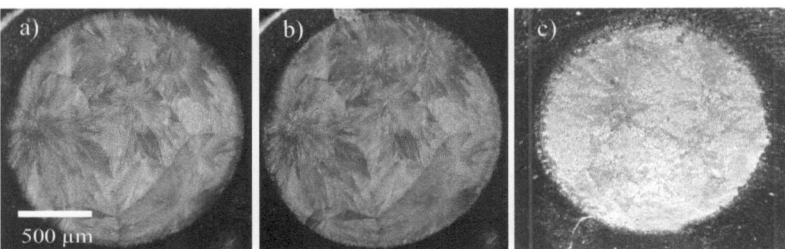

Figure 5.6 Images of temperature indicator in test channel: (a) before lamination, (b) after lamination (c) positive control (> 65°C).

To summarize, both temperature tests provide satisfying results regarding the heat impact on possible pre-stored bioreagents with pillar heights up to 20 μm (applicable for the spotting and drying) during lamination. However, for freeze-dried reagents (please refer to section 4.3.2), an increased chamber depth is proposed as the reagent pillars will most likely exceed 20 μm.

Bond Strength

The most important criteria for evaluating the efficacy of a sealing technology is the mechanical bond strength. For microfluidic devices, conventional methods like peel tests [150] do not sufficiently reflect the microfluidic functionality and environment. Consequently, they often give deluding results [138]. A possible method for testing the bond strength of microfluidic devices is by gradually increasing the pressure (e.g. with pressurized gas) inside a channel until delamination of the foil. This reflects the maximum allowable overpressure for a given geometry. In general, the total channel surface area as well as the distance between features, i.e. effectively bonded area, impacts on the delamination pressure. To quantify bond strength, test chips (not shown) featuring straight channels exhibiting comparable total channel area in respect to the amplification chip are milled with variable distances between channels and sealed based on optimal parameters. The minimal distance $d_{channels}$ amounts to 500 μm and increases up to 2 mm in 500 μm steps to enable the evaluation of delamination pressure in respect to bonded surface area. A connector pin is then glued to the inlets of a single channel in a channel pair. Subsequently, a pressure source is connected and the pressure is gradually

5.1 Temperature Diffusion Bonding

increased until delamination occurs. The results are summarized in figure 5.7. It can be seen that the bonded chips can withstand about 3 bars of overpressure even for only 500 µm wide bridges.

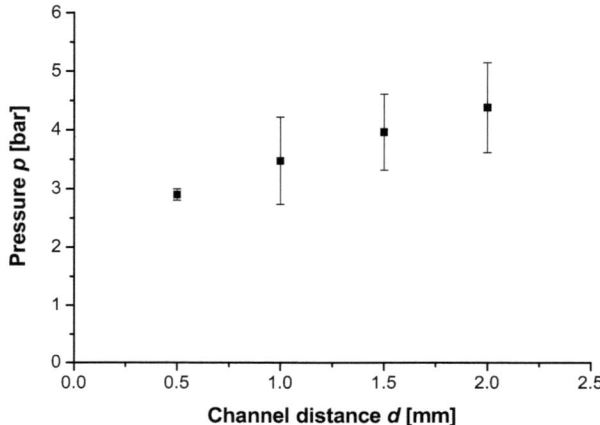

Figure 5.7 Delamination pressure in respect to distance between test channels. Already for 500 µm distance between channels, a strong bond can be observed.

Coating Compatibility

As discussed in section 3.2, a hydrophilic coating of substrate and potentially also the foil may be required for custom labs-on-a-chip and specifically for the amplification chip. Accordingly, substrate and foil are dip-coated with PEG (please refer to section 3.2.1). The PEG-coated chip and compound foil are then laminated with optimal parameters. In this case, no stable bond could be established. Even by increasing the roll temperature to 160°C which already caused deformations of the cover foil, no bonding to the substrate could be observed. This effect can be explained as the applied PEG coating (M_w of 15000 - 20000) melts at a temperature of 67.7°C [105] which is obviously lower than the temperature at the interface during the lamination. As a consequence, the coating acts as separating agent between substrate and cover foil (figure 5.8).

5.1 Temperature Diffusion Bonding

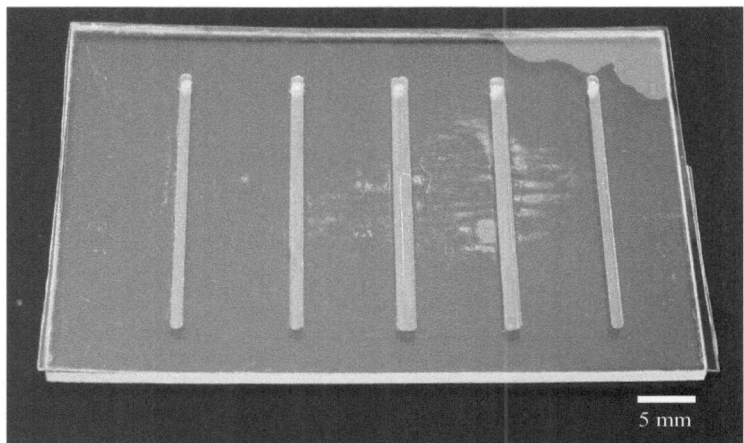

Figure 5.8 Image of temperature-diffusion-bonded, PEG-coated test chip. The melted PEG layer which promotes leakage is discernible between some channels.

A potential solution is the removal of the coating (or spill, depending on the means of coating) from the chip surface before bonding, e.g. with an alcohol-soaked wipe. A sealed chip which has been processed based on this method is depicted in figure 5.9. It demonstrates that even based on this procedure, a high quality bonding result can be achieved. However, this is no reproducible method and other solutions are desirable which will potentially be developed in follow-up work.

Optimized Process Chain

- Ultrasonicate chips for 30' in 2-propanol
- Dry with pressurized nitrogen
- Surface activation in O2-plasma, 4 min, 200 W
- Align chip & lid & protective foil in chip holder
- Move assembly 2 times through laminator at $v_{roll} = 0.4$ m/min, $T_{roll} = 140°C, p_{roll} = 2$ bar

5.1 Temperature Diffusion Bonding

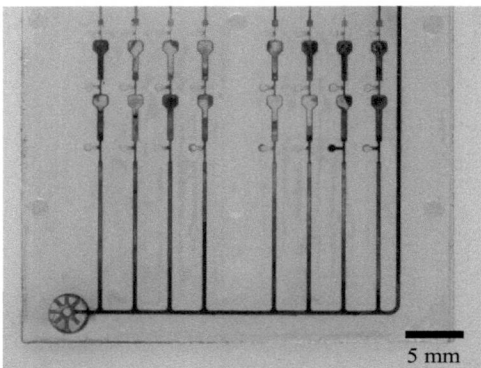

Figure 5.9 Example image of temperature-diffusion--bonded amplification chip primed with black dyed water for better visualization. All channels feature a PEG coating and no leakage is observable.

5.2 Adhesive Bonding

5.2.1 Principle

Adhesive bonding is based on the crosslinking of polymer chains in the liquid adhesive layer between bonding partners. The crosslinking or curing of adhesive can be initiated by different means, for example temperature or UV radiation. A good wetting of the bonding partners with adhesive is imperative as the bond is established in the surface roughness of the chip and lid. As the adhesive composition differs from that of the bonding partners, the fluid properties, optical properties and biocompatibility of the adhesive have to be taken into consideration. Still, adhesive bonding is the most uniform approach as it primarily relies on the wetting of the chip surface by the adhesive and is therefore applicable on various polymer materials. Thus, the following work investigates the influence of different process parameters for the adhesive bonding in more detail while trying to abstract the parameters from the utilized device and chip geometry.

A common approach for adhesive bonding is the coating of the lid with liquid adhesive. The lid is then aligned on the chip surface to achieve a bond. This way however, a large amount of (cured) adhesive gets in direct contact with reagents in the sample. Further, no bioreagents can be spotted on the lid (if applicable). Thus, different approaches to only selectively cover the chip surface with adhesive have been reported, namely contact printing [56, 61, 62, 65], the use of guide channels [60, 64] or a laminator [63]. The here applied technology is based on the latter approach, i.e. the transfer of an adhesive layer onto a microfluidic substrate *via* rolls (figure 5.10). Any contact between the adhesive layer and potentially spotted bioreagents on the lid can thus be prevented. When the substrate gets in contact with the rotating transfer roll, about half of the amount of adhesive present on the roll will be taken up by the chip. Thus, the process is independent of the chip features. The transferred amount of adhesive depends on the distance between the definition roll and the transfer roll, the viscosity of the adhesive as well as the transport velocity. For a fixed distance, the layer thickness can be extrapolated according to equation 5.1 [151].

$$d_{adh} \sim \frac{d_{roll}}{2}(\eta_{adh} \cdot v_{roll})^{0,64} \tag{5.1}$$

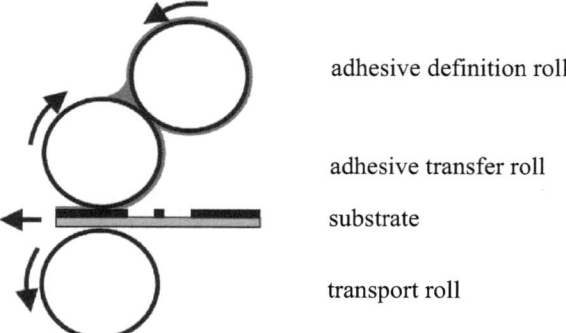

Figure 5.10 Principle for adhesive transfer. The amount of adhesive is controlled via the distance between the adhesive definition roll and the adhesive transfer roll. The transfer roll moves the chip through the laminator. Courtesy of NMI [152].

5.2.2 Materials and Methods

Microfluidic Test Chip

In contrast to the previous work, a different microfluidic chip has been used for process development. This can be explained by the unsuitable surface planarity of the injection-molded amplification chip as the master exhibits milling tracks of variable height (not shown). Therefore, a thicker (> 5 µm) adhesive layer is required to achieve good bonding results which however largely increases the risk of channel clogging. Thus, it has not been possible to bond amplification chips with acceptable yields based on this method.

The applied test chip with outer dimensions of standard microscope slides comprises 8 parallel channels with a sample volume of 7.5 µL each (figure 5.11). The reaction area features a width/depth of 500 µm while the supply channels exhibit a width/depth of 200 µm. The chips are currently fabricated by CNC-micromachining from COP [M17]. The chip assembly additionally comprises a lid (1mm thick, not shown).

5.2 Adhesive Bonding

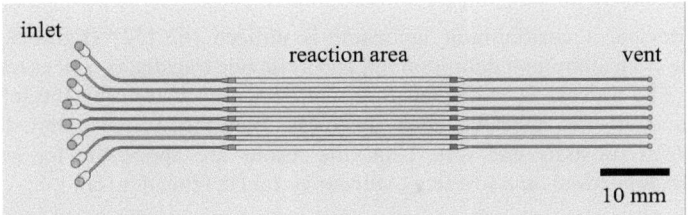

Figure 5.11 Microfluidic test chip featuring 8 parallel channels comprising inlets, reaction areas and vents.

Pre-Processing

Before bonding, the chip and lid are first ultrasonicated for 30' in 2-propanol to remove debris from the channels (please refer to section 3.1.1). This is followed by a drying step utilizing pressurized nitrogen. To enable adhesion, the adherent has to feature a higher or at least equal surface energy than the adhesive. Due to the low surface energy of polyolefines [153], a pretreatment of the surface is imperative for adhesive bonding. A common way of increasing the surface energy is oxygen plasma activation resulting in a hydrophilic surface. Thus, the surfaces to be bonded are activated in an oxygen plasma (please refer to section 3.1.3).

Adhesive

In adhesive bonding, reactive adhesives (2-component, e.g. epoxy) [56, 61, 62] and UV-curable adhesives [60, 63, 64, 65] are most frequently applied. However, the UV-radiation during bonding of the latter can possibly denaturate stored bioreagents if not masking the respective areas and preventing reflection or scattering. Thus, only epoxy-based adhesives which feature USP class VI approval (implantable, no cytotoxicity) [154] have been considered for this work. Specifically, a high T_g epoxy [M18] is selected which is also applicable for PCR-based applications and ensures no decrease in bond strength during cycling. Further, the high viscosity of the adhesive (~ 4 Pas) reduces the risk of channel clogging. To confirm the proposed bonding parameters, a less viscous adhesive (~ 1.2 Pas) [M19] is also evaluated.

5.2 Adhesive Bonding

Adhesive Transfer

As device, a custom-built laminator is utilized [63,152] (figure 5.12). It features an aluminum definition roll and a silicone transfer as well as transport roll. The process is abstracted from the device by measuring the influence parameters, i.e. adhesive layer thickness, transport velocity and distance between substrate and roll. Thus, the results are applicable for arbitrary laminators based on a soft (e.g., silicone or rubber) transfer roll.

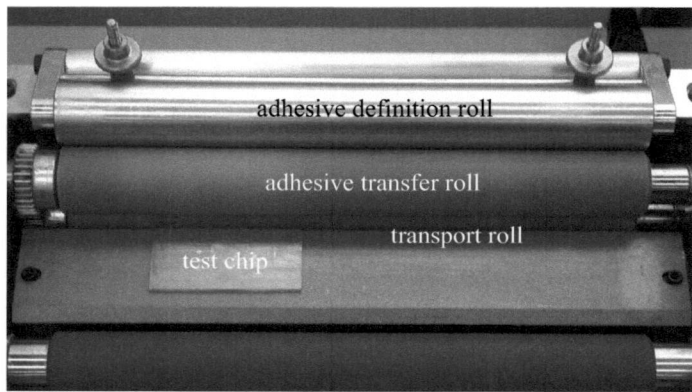

Figure 5.12 Modified laminator for adhesive transfer featuring an adhesive definition, adhesive transfer and transport roll.

Adhesive Layer Thickness

The amount of adhesive is measured gravimetrically [D14] and the layer thickness is extrapolated based on this data. For a good wetting (i.e. activated) surface, a homogenous distribution of adhesive over the chip surface can be assumed.

Post-Processing

After adhesive transfer, the chip is placed in an aluminum chip holder (not shown) comprising cavities for the chip/lid assembly. The depth of the cavities is 200 µm less than the total thickness of the assembly. Then, the lid is aligned on top of the chip. An aluminum plate is subsequently pressed on the holder via screw clamps [M20] which are able to exert a force > 1 kN. The chip holder is then evacuated at 10 mbar. Finally, the chip holder is put into a lab furnace [D23] for 3 h at 70°C for curing. It should be noted that the adhesive can also be cured at lower temperatures (e.g., 45°C) to prevent e.g. NASBA reagents from denaturating although cure times > 12 h are required. Thus, the higher cure temperature was selected during process development.

5.2.3 Evaluation

Adhesive Layer Thickness

With the selected adhesive and laminator, the minimum transferred layer thickness amounts to 3.5 µm. This can be explained by the incompressibility of the adhesive and the elasticity of the transfer roll. For achieving thinner layers, a larger amount is first transferred onto the chip surface followed by a second run through the laminator with no adhesive present thus effectively halving the amount of adhesive on the chip surface. Alternatively, an additional roll could be installed on the laminator. The results of the adhesive transfer exhibiting a good reproducibility (CV < 4 %) are summarized in figure 5.13. Further, it has been evaluated that the transferred layer thickness is basically independent of the feeding orientation of the chip into the laminator (tested for a given transfer roll velocity of 20 mm/s, not shown). Additionally, the same amount of adhesive is transferred onto the smaller features (500 µm wide spacers between the channels in the center chip area) as well as on the outer chip area. However, it should be noted that a small adhesive meniscus may be present at the interface between channel and lid.

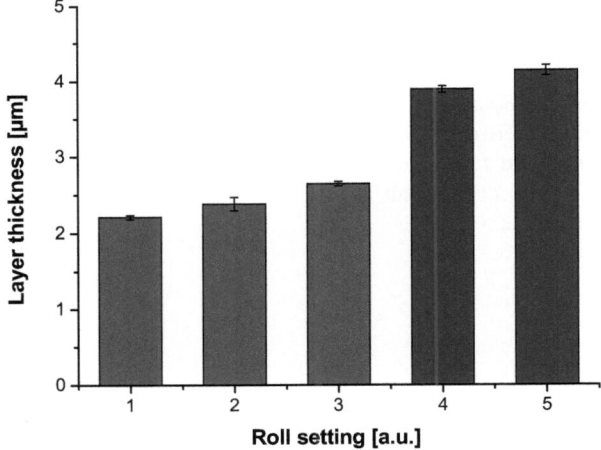

Figure 5.13 Extrapolated adhesive layer thicknesses. The first three values were achieved by moving adhesive-covered substrates through the laminator without any adhesive on the transfer roll.

Bonding Influence Parameters

For successfully bonding microfluidic chips, a high number of influence parameters have to be taken into account. The primary parameters for adhesive bonding, their influence and possible consequences are summarized in table 5.1 (in the order of processing). It should be noted that the chip design is the major influence parameter for bonding, i.e. chips featuring large (> 1 mm), deep (> 500 µm) and straight channels, only, will require less bonding process development due to the reduced risk of clogging than chips comprising small (< 500 µm), shallow (< 200 µm) and undulated channels as well as isolated features. The other parameters of major influence are evaluated in the following paragraph except for chip surface planarity and surface activation. The latter can be considered binary because either there is adhesion or not which depends on the wetting of the adhesive on the chip material. The chip surface planarity can be a major issue when using injection-molded chips (as is the case for the amplification chip) specifically if the master exhibits two layers of different height on the chip surface to facilitate fabrication. Then, the amount of transferred adhesive must be sufficient to compensate for the height difference. This however increases the risk of channel clogging.

For parameters of minor influence, i.e. rinsing, cleanliness of atmosphere as well as alignment relate to the processing in general or handling. A variation in thickness, if existent, requires the bonding partners to be in intimate contact during post-processing and cure. For ideal substrates (e.g., silicon wafers), the basic adhesion between chip, adhesive and lid may be sufficient to result in a strong bond. The following can be seen as a rule of thumb for adhesive bonding: the less ideal the surfaces of the bonding partners, the more pressure and/or adhesive is required.

5.2 Adhesive Bonding

Table 5.1 Legend:
++ major impact, i.e. a failure in bonding can occur
+ minor impact, can reduce yield
0 no impact.
* test parameter, please refer to table 5.1
** influences adhesive layer thickness [151]
*** increases risk of clogging shallow (< 100 μm) features.)

Property	Influence	Consequence
Chip surface planarity	++	More adhesive required
Chip thickness variation	+	Higher pressure required
Rinsing	+	Residual fat prevents adhesion
Surface activation	++	No wetting of chip surface
Layer thickness	++	No bond vs. clogging
Transport velocity	0*	
Distance substrate-roll	0**	
Cleanliness of atmosphere	+	Dirt particles, local leaks
Alignment	+	Smearing of adhesive into channels
Applied pressure	++	No bond
Evacuation	++	Leaks around small features
Cure	0	

Chip Bonding Results

In general, successfully bonded chips do not exhibit any leakage even if operated for 2 h at 100 °C with an overpressure of 4 bar. Further, no cross-flow between channels could be observed (12 h @ 4 bar) when alternatingly priming the channels with dyed liquid and DI-water. For the presented test chip, the different microfluidic channels can be seen as independent units, i.e. clogging of one channel does not result in chip failure. Thus, the results of the bonding have been classified in the following way:

5.2 Adhesive Bonding

- Fail: multiple leaks and/or > 2 channels clogged
- Working: no internal leaks, up to two channels clogged
- Good: no leaks, one channel clogged
- Perfect: no leaks, all channels operational

The results of the chip bonding are summarized in table 5.2 First, it is imperative to apply a high pressure on bonding partners during post-processing and cure. Second, for the presented chip geometry, no good bonding results could be achieved without an evacuation step after the adhesive transfer. The evacuation step can lead to a redistribution of adhesive on the interface between chip and lid and thus can compensate for process variations. It should however be noted that a very long evacuation leads to clogging as the adhesive is also pulled into the channels over time. Further, transferring ~ 2.5 µm onto the chip surface is the recommended layer thickness for the selected adhesive. A higher amount results in more clogged channels (figure 5.15-a) while a smaller amount does not exhibit sufficient cohesive forces to result in a strong bond.

Table 5.2 Bonding results for different process parameters. If not stated otherwise, adhesive [M18] layers with d_{adh} ~ 2.5 µm are transferred, the chip and lid assembly are evacuated for t_{vac} = 3 h under constant pressure of ~ 1 kN and cured directly afterwards for t_{cure} = 3 h. Best results are achieved for chips featuring capture channels. The last two rows refer to experiments based on [M19] as adhesive in combination with capture channels.

Process Variation	Perfect	Good	Working	Fail
No pressure@ cure	0	0	0	3
t_{vac} < 2 h	0	0	1	7
t_{vac} > 12 h	0	0	0	3
d_{adh} < 2 µm	0	0	0	5
d_{adh} > 3 µm	0	0	11	3
d_{adh} ~ 2.5 µm	0	0	5	0
t_{store} = 48 h	0	3	0	0
Remove adhesive	0	3	0	0
Capture channels	35	2	3	0
d_{adh} < 2 µm [M19]	1	0	0	5
d_{adh} ~2.5 µm [M19]	10	0	0	2

5.2 Adhesive Bonding

Still, even based on optimal parameters, it has not been possible to achieve a "perfect" bonding result (figure 5.15-b). The results can be improved by storing the chips at room temperature after evacuation. This however reduces the throughput and thus alternate ways are desirable. It has been observed that for the presented chip, only the outer channels tend to clog in most cases. It is thus assumed that the distance between a feature and the next larger, adhesive-covered area is the prime factor for clogging of channels. Some excess adhesive has to be present on the chip surface to compensate for e.g. a variation in surface planarity. Adjacent channels can thus be clogged if near to a larger area covered with adhesive, i.e. near a supply.

To test this assumption, the adhesive has been manually removed from the chip sides after transfer with an acetone wipe which again leads to improved results. A more defined approach however is the integration of capture channels which collect excess adhesive and thus prevent the outer channels from clogging (figure 5.14, figure 5.15-c).

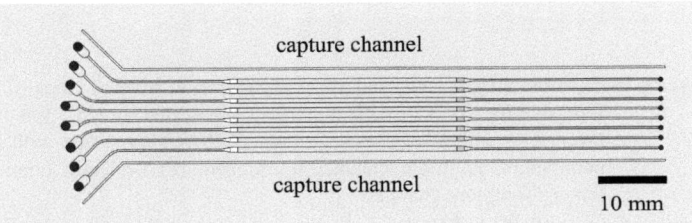

Figure 5.14 Microfluidic test chip featuring 8 parallel channels comprising inlets, reaction areas and vents.

For the applied test chips, 200 µm wide and deep channels are milled in a distance of 1 mm to the respective outer reaction area. The dimensions of the capture channels should be equal or smaller than the shallowest or least wide functional channel. Otherwise, excess adhesive would flow into channels exhibiting the highest capillary pressure and not into the capture channels.Thus, it is assumed that even higher yields (table 5.5) should be achievable by optimizing the position of the capture channels. For the less viscous adhesive [M19], about the same layer thickness has to be transferred in respect to the primarily evaluated adhesive [M18] to achieve perfect bonding results based on chips featuring capture channels. The failed chips correlate to the increased risk of channel clogging.

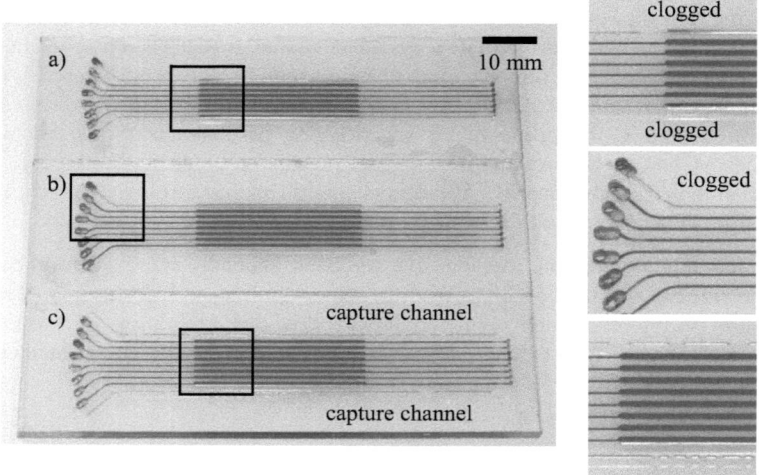

Figure 5.15 Example images (left) and close-ups (right) of bonded chips filled with food dye for better visualization. (a) Chip sealed by a 3.5 μm adhesive layer. The outer channels are clogged. (b) Chips sealed with optimal parameters. A single channel is clogged. (c) Perfectly bonded chip featuring capture channels.

Coating Compatibility: PEG

Inherently, depending on the cure protocol, adhesive bonding does not exhibit an incompatibility to PEG coated surfaces in contrast to the temperature diffusion bonding. However, several tests have shown that bond the strength is reduced (< 2 bar for the amplification chip, please refer to section 5.1.3) when bonding PEG-coated surfaces. An additional plasma activation step of only the lid or foil however leads to a strong bond. Furthermore, substrate and cover foil have to be cleaned with 2-propanol after PEG coating. The latter can be explained by the removal of excess polymer chains which are only loosely bound on the surface. Water contact angles [D6] of differently modified test samples of COC and COP are summarized in figure 5.16. After plasma activation (please refer to section 3.1.3), the surfaces exhibit decreased hydrophilicity compared to PEG coated surfaces.

5.2 Adhesive Bonding

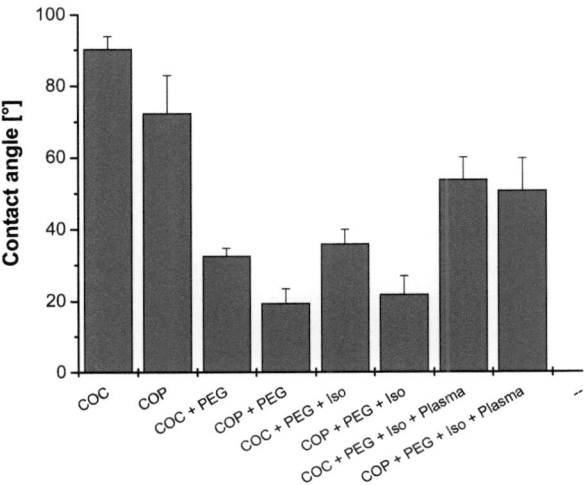

Figure 5.16 Contact angle measurements of differently processed COC and COP surfaces. Plasma activation of PEG-covered surfaces leads to a decrease in hydrophilicity.

To further investigate this effect, X-ray photoelectron spectroscopy (XPS) [D24] measurements are conducted for COC-based test samples. Plain COC does not comprise atomic oxygen however PEG does. Thus, the ratio between the C1s and O1s electrons [156] is evaluated (table 5.3, survey scan in figure 5.17). In contrast to the plain sample, an increase in O1s electrons is observable. Still, it can not be excluded that the plasma process itself completely removes the PEG layer and the resulting O1s electrons are detected due to the plasma process [157] after coating. Thus, the NASBA or assay compatibility of this process has to be verified which has not been possible due to time constraints.

5.2 Adhesive Bonding

Table 5.3 Ratio of C1s and O1s electron counts of differently processed COC surfaces measured *via* XPS.

Processing	C1s	O1s
Plain	98.54	1.46
Oxygen plasma	66.7	33.3
PEG	68.64	31.36
PEG + 2-propanol cleaning	69.49	30.51
PEG + plasma	83.29	16.71

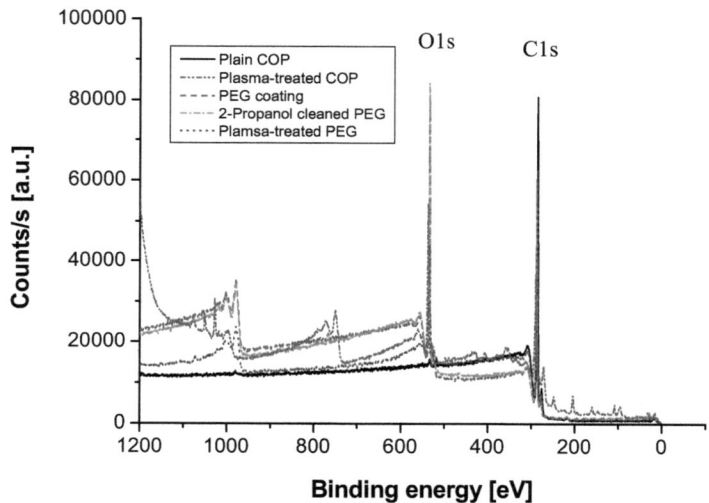

Figure 5.17 XPS survey scan of differently processed COC surfaces exhibiting more or less prominent peaks for O1s and C1s electrons.

Coating Compatibility: Vistex

The PCR-compatible coating (please refer to section 3.2.5) does exhibit very good compatibility to adhesive bonding based on epoxy due to formation of amide bonds [155] between coating and adhesive. Accordingly, similar bonding results could be achieved for coated as well as non-coated chips.

5.2 Adhesive Bonding

Optimized Process Chain

- Ultrasonicate chips for 30' in 2-propanol
- Dry with pressurized nitrogen
- Surface activation in O_2-plasma, 4 min, 200 W
- Transfer 2.5 µm of Epo-Tek 375 onto chip surface
- Align chip & lid in chip holder
- Apply force > 1 kN on chip assemblies
- Evacuate assembly for 3h @ 10 mbar
- Cure assembly for 3h @ 70°C

5.3 Conclusion & Outlook

In this chapter, two methods for the biocompatible sealing of polymer chips have been presented and evaluated, namely temperature diffusion bonding and adhesive bonding.

Specifically, existing work by Steigert et al. [59] for temperature diffusion bonding has been extended to allow for optical read-out techniques based on the novel, transparent compound foil. The adapted process further results in strongly bonded test chips. A potentially negative temperature impact of the sealing process on pre-stored reagents, e.g. enzymes has been excluded. Finally, it has been evaluated that PEG as hydrophilic coating poses a risk in combination with temperature diffusion bonding as residual PEG on the chip surface promotes delamination. Consequently, it is proposed to prevent any PEG residue on the chip surface or to remove spill before sealing.

For adhesive bonding, relevant process influence parameters have been evaluated. Consequently, a process chain with parameter recommendations is provided which can greatly reduce the time and costs for sealing development of a custom lab-on-a-chip. The process is inherently compatible with Vistex as high-temperature hydrophilic coating. Further, the bonding principle in combination with the selected adhesive allows for applications where high temperature stability and reagent storage on-foil is required, e.g. a solid-phase PCR.

To summarize, both methods are capable of meeting the requirements for sealing of polymer labs-on-a-chip. Still, more work has to be invested in respect to PEG compatibility which possibly requires adjustments of the processes.

Chapter 6
General Conclusions and Outlook

This thesis reported about back-end processing of polymer labs-on-a-chip. Multiple processing steps for surface modification, reagent pre-storage and biocompatible sealing have been discussed and validated by proof-of-principle experiments. In total, these processing steps provide means to realize complex applications like biological assays on polymer-based labs-on-a-chip.

Specifically, approaches for the surface modification of polymer labs-on-a-chip ranging from cleaning or activation steps, hydrophilization to selective hydrophobization have been presented. Especially the latter allows for flexible fabrication of strong and reliable hydrophobic valves. The novel superhydrophobic coating, namely Teflon-carbon black, could potentially be applicable in completely different scenarios due to its simplicity and low cost, especially when comparing it to clean-room based approaches to achieve superhydrophobicity. For labs-on-a-chip, its biocompatibility should permit application in a wide range of assays.

Two approaches for the pre-storage of dry reagents which is a prime requirement for fully integrated labs-on-a-chip have been presented and evaluated. It has been demonstrated that spotting and drying is a feasible concept for the storage of DNA/RNA primers as well as probe molecules, i.e. labelled oligos, by conducting NASBA after long-term storage of respective reagents. This concept however is not viable for long-term storage of susceptible enzymes. Accordingly, a protocol for the freeze-drying of reagents on-chip has been developed which allows for extended storage as well as position-specific, sub-microliter reagent deposition.

For biocompatible sealing of polymer substrates, two methods featuring high bond strength have been presented and evaluated, namely temperature diffusion bonding and adhesive bonding. Both processes permit pre-storage of reagents on-chip either due to acceptable temperature impact or due to inherent biocompatibility. The work based on temperature diffusion bonding allows for optical read-out techniques due to the introduced transparent compound foil. An in-depth investigation on relevant influence parameters for adhesive bonding as universal sealing approach has been conducted. A resulting processing chain with parameter recommendations to reduce time and costs for sealing development has been provided. For both sealing methods, the compatibility to potential hydrophilic coatings has to be taken into special consideration. It has been determined that temperature diffusion bonding is especially susceptible to coating residue. Thus, it is proposed to apply coatings which are inherently compatible with the sealing process or to prevent any spill on the bonding interface.

To summarize, a mature back-end process chain for arbitrary polymer labs-on-a-chip is provided which can be seen as a toolbox to facilitate development and reduce time and thus costs. Still, no universal approach can be proposed as the selection of suitable processes is primarily dependent on the given boundary conditions for development. Especially for alternate substrate materials and coatings, different approaches are more or less feasible. Nonetheless, it was the aim of this thesis to provide means for different application scenarios even if most of the evaluation was based on a single lab-on-a-chip. Further, critical issues have been pointed out which reflect the complexity of back-end chip processing. It is thus again emphasized that not only the fluidic concept or chip fabrication plays an important role in lab-on-a-chip development but also the back-end processing which is mostly taken for granted.

Future work should mainly focus on providing more suitable combinations of substrate material, hydrophilic coating and sealing technique as this poses the greatest challenge. Here, application scenarios for low temperatures like immunoassays or NASBA may require a completely different approach than scenarios for high temperatures like PCR. When moving from prototyping to pre-production and thus commercialization, the focus will further have to shift to scalability of processing or throughput.

Chapter 7
References

7.1 Literature

[1] P. Yager, G. J. Domingo and J. Gerdes, Point-of-Care Diagnostics for Global Health, *Annual Review of Biomedical Engineering*, **10**, pp. 107-144, 2008

[2] N. Drenck, Point of care testing in Critical Care Medicine: the clinician's view, *Clinica Chimica Acta*, **307**, pp. 3-7, 2001.

[3] C. Price, Point of care testing, *BMJ*, **322**, pp. 1285–1288, 2001.

[4] C. P. Price, Point of Care Testing: Potential for Tracking Disease Management Outcomes, *Disease Management & Health Outcomes*, Vol 10, **12**, pp. 749-761, 2002

[5] P. St.Louis, Status of Point-of-Care Testing: Promise, Realities, and Possibilities, *Clinical Biochemistry*, Vol. 33, **6**, pp. 427–440, 2000.

[6] P. Lode, Point-of-care immunotesting: Approaching the analytical performance of central laboratory methods, *Clinical biochemistry*, Vol. 38, **7**, pp. 591-606, 2005.

[7] K. A. Erickson and P. Wilding, Evaluation of a novel point-of-care system, the i-STAT portable clinical analyzer, *Clinical Chemistry*, Vol 39, 283-287, 1993.

[8] M. Mueller-Bardorff, T. Rauscher, M. Kampmann, S. Schoolmann, Quantitative Bedside Assay for Cardiac Troponin T: A Complementary Method to Centralized Laboratory Testing, *Clinical Chemistry*, 45, **7**, pp. 1002-1008, 1999.

[9] Apple, R. Christenson, R. Valdes, A. Andriak, Simultaneous Rapid Measurement of Whole Blood Myoglobin, Creatine Kinase MB, and Cardiac Troponin I by the Triage Cardiac Panel for Detection of Myocardial Infarction, *Clinical Chemistry*, 45, **2**, pp. 199-205, 1999.

[10] R. F. Louie, Z. Tang, D. V. Sutton, J. H. Lee, G. J. Kost, Point-of-Care Glucose Testing, *Archives of Pathology & Laboratory Medicine*, Vol. 124, **2**, pp. 257-266, 2000.

[11] Wu, A. Smith, R. Christenson, M. Murakami, F. Apple, Evaluation of a point-of-care assay for cardiac markers for patients suspected of acute myocardial infarction, Clinica Chimica Acta, 346, pp. 211–219, 2004.

[12] C. H. Ahn, J.-W, Choi, G. Beaucage, J. H. Nevin, Disposable Smart Lab

on a Chip for Point-of-Care Clinical Diagnostics, Proceedings of the IEEE, 92-1, 154-173, 2004.

[13] D. Figeys, D. Pinto, Lab-on-a-chip: A revolution in biological and medical sciences, *Analytical Chemistry*, 72, **9**, pp. 330A–335A, 2000.

[14] T. Thorsen, S. J. Maerkl, S. R. Quake, Microfluidic large-scale integration, *Science*, 298, pp. 580–584, 2002.

[15] D. Erickson, D. Q. Li, Integrated microfluidic devices, *Analytica Chimica Acta*, 507, **1**, pp. 11–26, 2004.

[16] T. Schulte, R. Bardell, B. Weigl, Microfluidic Technologies in Clinical Diagnostics, *Clinica Chimica Acta*, pp. 1–10, 2005.

[17] D. J. Harrison, C. Wang, P. Thibeault, F. Ouchen, S. Cheng, The Decade Search for the Killer App in µTas, Kluwer Academic Publisher, pp. 195–204, 2000.

[18] D. Figeys and D. Pinto, Lab-on-a-chip: A Revolution in Biological and Medical Sciences, 2000, Anal. Chem., 72(9), 330A-335A

[19] D. Reyes, D. Iossifidis, P. Auroux, A. Manz, Micro total analysis systems. Part 1. Introduction, theory, and technology, *Analytical Chemistry*, 74, **12**, 2623–2636, 2002.

[20] P. A. Auroux, D. Iossifidis, D. R. Reyes, A. Manz, Micro total analysis systems: 2 Analytical standard operations and applications. *Analytical Chemistry*, 74, **12**, pp. 2637–2652, 2002.

[21] R. E. Oosterbroek, A. van den Berg, Lab-on-a-Chip: Miniaturized systems for (bio)chemical analysis and synthesis, Elsevier Science: Amsterdam, NL, 2003.

[22] T. Vilkner, D. Janasek, A. Manz, Micro total analysis systems. Recent developments, *Analytical Chemistry*, 76 (12), pp. 3373–3385, 2004.

[23] S. Haeberle and R. Zengerle, Microfluidic Platforms For Lab-On-A-Chip Applications, *Lab Chip*, 7, 1094 – 1110, 2007.

[24] D. Mark, S. Haeberle, G. Roth, F. von Stetten, R. Zengerle, Microfluidic Lab-on-a-Chip Platforms: Requirements, Characteristics and Applications, *Chem. Soc. Rev.*, DOI:10.1039/b820557b, 2010.

[25] T. Baier, T. E. Hansen-Hagge, R. Gransee, A. Crombé, S. Schmahl, C. Paulus, K. S. Drese, H. Keegan, C. Martin, J. J. O'Leary, L. Furuberg, L. Solli, P. Grønn, I. M. Falang, A. Karlgård, A. Gulliksen, F. Karlsen, Hands-free sample preparation platform for nucleic acid analysis, *Lab on a Chip*, **9**, pp. 3399 - 3405, 2009.

[26] L. Riegger, M. Grumann, J. Steigert, S. Lutz, C. P. Steinert, C. Mueller,

J. Viertel, O. Prucker, J. Ruehe, R. Zengerle and J. Ducrée, Single-step Centrifugal Hematocrit Determination on a 10-$ Processing Device. *Biomed Microdevices*, **9**, pp. 795-799, 2007.

[27] M. Grumann, A. Geipel, L. Riegger, R. Zengerle and J. Ducreé, Batch-mode mixing on centrifugal microfluidic platforms, *Lab Chip*, 2005, 5, **5**, 560–565.

[28] L. Riegger, M. Grumann, T. Nann, J. Riegler, O. Ehlert, W. Bessler, K. Mittenbühler, G. Urban, L. Pastewka, T. Brenner, R. Zengerle, J. Ducrée, Read-out concepts for multiplexed bead-based fluorescence immunoassays on centrifugal microfluidic platforms, *Sensors Actuators A*, **126**, pp. 455-462, 2006.

[29] T. Lee, M. Carles, I-M. Hsing, Microfabricated PCR-electrochemical device for simultaneous DNA amplification and detection, *Lab on a Chip*, **3**, pp. 100–105, 2003.

[30] S. Sia, V. Linder, B. Parviz, A. Siegel, G. Whitesides, An integrated approach to a portable and low-cost immunoassay for resource-poor settings, *Angew. Chem. Int. Ed.*, **43**, pp. 498-502, 2004.

[31] J. Gao, X.-F. Yin, Z.-L. Fang, Integration of single cell injection, cell lysis, separation and detection of intracellular constituents on a microfluidic chip, *Lab Chip*, **3**, pp. 47-52, 2004.

[32] Y. K. Cho, J. G. Lee, J. M. Park, B. S. Lee, Y. Lee, and C. Ko, "One-step pathogen specific DNA extraction from whole blood on a centrifugal microfluidic device, *Lab on a Chip*, Vol. 7, **5**, pp. 565-573, 2007.

[33] R. H. Liu, J. Yang, R. Lenigk, J. Bonanno and P. Grodzinski, Self contained, fully integrated biochip used for sample preparation, polymerase chain reaction amplification and DNA microarray detection, *Analytical Chemistry*, 76, **7**, pp 1824–1831, 2004.

[34] L. Kang-Yi, L. Wang-Ying, L. Yu-Fang, W. Chih-Hao, L. Huan-Yao, L. Gwo-Bin, Microfluidic System Integrated with a sample pre-treatment device for fast nucleic acid amplification, *Journal of microelectromechanical systems*, Vol. 17, **2**, pp. 288-301, 2008

[35] A. F. Sauer-Budge, P. Mirer, A. Chatterjee, C. M. Klapperich, D. Chargin and A. Sharon, Low cost and manufacturable complete microTAS for detecting bacteria, *Lab Chip*, **9**, pp. 2803-2810, 2009.

[36] D. Chen, M. Mauk, X. Qiu, C. Liu, J. Kim, S. Ramprasad, S. Ongagna, W. R. Abrams, D. Malamud, P. L. A. M. Corstjens and H. H. Bau1, An integrated, self-contained microfluidic cassette for isolation,

amplification, and detection of nucleic acids, DOI 10.1007/s10544-010-9423-4.

[37] P. Yager, T. Edwards, E. Ful, K. Helton, K. Nelson, M. R. Tam and B. H. Weigl, Microfluidic diagnostic technologies for global public health, *Nature*, **442**, pp. 412-418, 2006.

[38] H. Becker and C. Gaertner, Polymer microfabrication technologies for microfluidic systems, *Anal Bioanal Chem*, **390**, pp. 89–111, 2008.

[39] F. Walther, P. Davydovskaya, S. Zuercher, M. Kaiser, H. Herberg, A. M. Gigler and R. W. Stark, Stability of the hydrophilic behavior of oxygen plasma activated SU-8, *J. Micromech. Microeng.*, **17**, pp. 524–531, 2007.

[40] Y. Yuan, C. S. Liu, Y. Zhang1, M. Yin, Jie Xu, Effect of Oxygen on Surface Properties and Drug Release Behavior of Plasma Polymer of n-Butyl Methacrylate, *Chinese Chemical Letters*, Vol. 16, **12**, pp 1641-1644, 2005.

[41] B. Chang et al., Surface-Attached Polymer Monolayers for the Control of Endothelial Cell Growth, C*olloids and Surfaces A: Physicochemical and Engineering Aspects*, 198-200, pp. 519-526, 2002.

[42] J. D. Cox, M. S. Curry, S. K. Skirboll, P. L Gourley, D. Y. Sasaki, Surface passivation of a microfluidic device to glial cell adhesion: a comparison of hydrophobic and hydrophilic SAM coatings, *Biomaterials*, **23**, pp. 929–935, 2002.

[43] K. Handique, B. P. Gogoi, D. T. Burke, C. H. Mastrangelo, and M. A. Burns, Microfluidic flow control using selective hydrophobic patterning, *Proc. SPIE*, Vol. 3224, pp 185-187, 1997.

[44] K. Handique, D. T. Burke, C. H. Mastrangelo, and M. A. Burns, Nanoliter Liquid Metering in Microchannels Using Hydrophobic Patterns ,*Anal. Chem.*, **72**, pp 4100-4109, 2000.

[45] J. W. Suk and J. H. Cho, Capillary flow control using hydrophobic patterns, *J. Micromech. Microeng.*, **17**, N11–N15, 2007.

[46] A. Gliére, C. Delattre, Modeling and fabrication of capillary stop valves for planar microfluidic systems, *Sensors and Actuators A*, **130–131**, pp 601–608, 2006.

[47] A. Puntambekar, J.-W. Choi, C. H. Ahn, S. Kim and V. Makhijani, Fixed-volume metering microdispenser module, *Lab Chip*, **2**, pp 213–218, 2002.

[48] S.-H. Lee, C.-S. Lee, B.-G. Kim, Y.-K. Kim, Quantitatively controlled

nanoliter liquid manipulation using hydrophobic valving and control of surface wettability, *J. Microm. Microeng.*, **13**, pp. 89-97, 2003.

[49] P. Andersson, G. Jesson, G. Kylberg, G. Ekstrand, and G. Thorsen, Parallel Nanoliter Microfluidic Analysis System, *Anal. Chem.*, **79**, pp. 4022-4030, 2007.

[50] Gyrolab Bioaffy, Gyros AB, Sweden, www.gyros.com, accessed 2010.

[51] F. Franks, Freeze-drying of bioproducts: putting principles into practice, European journal of Pharmaceutics and Biopharmaceutics, **45**, pp. 221-229, 1998.

[52] M. Brivio, Y. Li, A. Ahlford, B. G. Kjeldsen, J. L. Reimers, M. Bu, A.-C. Syvänen, D. D.Bang and A. Wolff, A Simple and Efficient Method for on-Chip Storage of Reagents:Towards Lab-on-a-Chip Systems for Point-of-Care Diagnostics, *Proc. Microtas*, pp. 59-61, 2007.

[53] C.-W. Tsao, D. L. DeVoe, Bonding of Thernoplastic Polymer Microfluidics, *Microfluid Nanofluid*, **6**, pp. 1-16, 2009.

[54] D. Erickson and D. Li, Integrated microfluidic devices, *Analytica Chimica Acta*, Vol. 507, **1**, pp. 11-26, 2004.

[55] C. M. Niemeyer, M. Adler and R. Wacker, Detecting antigens by quantitative immuno-PCR, *Nature Protocols*, **2**, pp.1918 - 1930, 2007.

[56] W. Eberhardt, H. Kueck, P. Koltay, M. Muench, H. Sandmaier, M. Spritzendorfer, R. Steger, M. Willmann, R. Zengerle, Low Cost Fabrication Technology for Microfluidic Devices Based on Micro Injection Moulding, Proc. Micro.tec, pp. 129 – 134, 2003.

[57] F. Umbrecht, D. Müller, F. Gattiker, C.M. Boutry, J. Neuenschwander, U. Sennhauser, Ch. Hierold, Solvent assisted bonding of polymethylmethacrylate: Characterization using the response surface methodology, Sensors and Actuators A 156, pp. 121–128, 2009.

[58] K. C.-H. Ahn, J.-W. Choi, G. Beaucage, J. H. Nevin, J.-B. Lee, A. Puntambekar, and J. Y. Lee, Disposable Smart Lab on a Chip for Point-of-Care Clinical Diagnostics, Proc. IEEE, Vol. 92, N. 1, 2004.

[59] J. Steigert, S. Haeberle, T. Brenner, C. Mueller, C. P. Steinert, P. Koltay, N. Gottschlich, H. Reinecke, J. Ruehe, R. Zengerle and J. Ducrée, *Rapid prototyping of microfluidic chips in COC*, J. Micromech. Microeng., **17**, pp. 333–341, 2007.

[60] Z. Huang, J. C. Sanders, C. Dunsmor, H. Ahmadzadeh, J. P. Launders, A Method for UV-bonding in the fabrication of glass electrophoretic microchips, Electrophoresis, **22**, pp. 3924-3929, 2001.

[61] F. Dang, S. Shinohara, O. Tabata, Y. Yamaoka, M. Kurokawa, Y. Shinohara, M. Ishikawa and Y. Baba, *Replica multichannel polymer chips with a network of sacrificial channels sealed by adhesive printing method*, Lab on a Chip, **5**, pp. 472-478, 2005.

[62] B. R. Flachsbart, K. Wong, J. M. I., E. N. Abante, R. L. Vlach, P. A. Rauchfuss, P. W. Bohn, J. V. Sweedler and M. A. Shannon, Design and fabrication of a multilayered polymer microfluidic chip with nanofluidic interconnects via adhesive contact printing, Lab on a Chip, **6**, pp. 667–674, 2006.

[63] J. Kentsch, S. Breitsch, M. Stelzle, Low temperature adhesion bonding for BioMEMS, J. Micromech. Microeng., **16**, pp. 802-807, 2006.

[64] C. Lu, L. J. Lee, Y.-J. Juang, Packaging of microfluidic chips via interstitial bonding technique, Electrophoresis, **29**, pp. 1407–1414, 2008.

[65] S. Carroll, M. M. Crain, J. F. Naber, R. S. Keynton, K. M. Walsh and R. P. Baldwin, Room temperature UV adhesive bonding of CE devices, Lab on a Chip, **8**, pp. 1564-1569, 2008.

[66] W. Adamson and A. P. Gast, *Physical chemistry of surfaces*, John Wiley & Sons, 6th edition, 1997.

[67] A. Frohn, N. Roth, *Dynamics of Droplets*, Springer-Verlag Berlin, 2000.

[68] G. Schramm, *Einführung in die Rheologie und Rheometrie*, Thermo Haake, Karlsruhe, 2. Aufl., 2002.

[69] H. Bruus, *Theoretical microfluidics,* Oxford University Press, Oxford, 1st ed., 2008.

[70] P. Roach, N. J. Shirtcliffe, M. I. Newton, Progress in superhydrophobic surface development, Soft Matter, **4**, pp. 224-240, 2008.

[71] C. Dorrer and J. Ruehe, Some thoughts on superhydrophobic wetting, *Soft Matter*, **5**, pp. 51-61, 2009.

[72] A. B. D. Cassie; S. Baxter, Trans. Faraday Soc., **40**, pp 546–551, 1944.

[73] D. Langbein, Capillary surfaces: shape, stability, dynamics, in particular under weightlessness, Springer, Berlin, 1st ed., 2002.

[74] L. Furuberg, M. M. Mielnik, A. Gulliksen, L. Solli, I.-R. Johansen, J. Voitel, T. Baier, L. Riegger, F. Karlsen, RNA amplification chip with parallel microchannels and droplet positioning using capillary valves, *Microsystem Technologies*, **14**, pp 673 - 681, 2008.

[75] SINTEF, Norway, www.sintef.no, accessed 2010.

[76] IMM, Institute of Microtechnology Mainz, Germany, www.imm-mainz.de, accessed 2010.

[77] NorChip, Norway, www.norchip.com, accessed 2010.

[78] BioFluidix, Germany, www.biofluidix.de, accessed 2010.

[79] Coombe Women's Hospital, Ireland, www.coombe.ie, accessed 2010.

[80] P. Pisani, F. Bray, D. M. Parkin, Estimates of the worldwide prevalence of cancer for 25 sites in adult population, *Int. J. Cancer*, **97**, pp. 72-81, 2002.

[81] F. X. Bosch, A. Lorincz, N. Munoz, C. J. L. M. Meijer, K. V. Shah, The causal relation between human papillomavirus and cervical cancer, *J. Clin. Pathol.*, **55**, pp. 244-265, 2002.

[82] N. Munoz, F. X. Bosch, S. de Sanjosé, R. Herrero, X. Castellsagué, K. V. Shah, P. J. F. Snijders, C. J. L. M. Meijer, for the International Agency for Research on Cancer, Multicenter Cervial Cancer Study Group, Epidermiologic classification of human papillomavirus types associated with cervical cancer, *N. Engl. J. Med.*, **384**, pp. 518-527, 2003.

[83] E. M. de Villiers, C. Fauquet, T. R. Broker, H. U. Bernard, H. zur Hausen, Classification of papillomaviruses, *Virology*, **324**, pp. 17-27, 2004.

[84] A. Clad, Cervixdysplasien (CIN), presentation slides, 2008.

[85] M. A. Nobbenhuis, T. J. Helmerhorst, A. J. van den Brule, L. Rozendaal, F. J. Voorhorst, P. D Bezemer, R. H. Verheijen, C. J. Meijer, Cytological regression and clearance of high-risk human papillomavirus in women with an abnormal cervical smear, *Lancet*, **358**, pp.1782-3, 2001.

[86] D. R. Lowy, P. M. Howley, D. M. Knipe, P. M. Howley, D. E. Griffin, R. A. Lamb, Papillomaviruses, Lippincott Williams & Wilkins, pp. 2231-2264, 2001.

[87] World Health Organization, Human papillomavirus, www.who.int, accessed 2010.

[88] K. Kühndel, U. Hettmer, Ch. Biesold, U. Köhler, Häufigkeit falsch negativer Abstriche bei Präkanzerosen und Frühstadien des Zervixkarzinoms, *Verh. Dtsch. Ges. Zyt.* **18**, pp.125-126, 1993.

[89] Hybrid Capture 2, Digene, www.digene.com, accessed 2010.

[90] J. Möckel, A. Clad, J. Quaas, H. Meisel, A. Endres, H. Heyer, R. Kirchmayr, J. Zimmermann, V. Schneider, E6/E7 mRNA-transcripts as

early indicators of progressive intraepithelial neoplasia of the cervix in a clinical setting, not yet published.

[91] K. Münger, W. C. Phelps, V. Bubb, P. M. Howley, R. Schlegel, The E6 and E7 genes of human papillomavirus type 16 together are necessary and sufficient for transformation of primary human keratinocytes, *J Virol*, **63**, pp. 4417-21, 1989.

[92] C. Rosty, M. Sheffer, D. Tsafrir, N. Stransky, I. Tsafrir, M. Peter, P. de Crémoux, A. de La Rochefordiére, R. Salmon, T. Dorval, JP. Thiery, Identification of a proliferation gene cluster associated with HPV E6/E7 expression level and viral DNA load in invasive cervical carcinoma, *Oncogene*, **24**, pp. 7094-104, 2005.

[93] G. Clifford, JS Smith, T. Aguado, S. Franceschi, Comparison of HPV type distribution in high-grade cervical lesions and cervical cancer worldwide: a meta-analysis, *BR J Cancer*, **88**, pp. 63-73, 2003.

[94] J. Compton, Nucleic acid sequence-based amplification, *Nature*, **350**, pp. 91-92, 1991.

[95] I. K. Dimov, J. L. Garcia-Cordero, J. O'Grady, C. R. Poulsen, C. Viguier, L. Kent, P. Daly, B. Lincoln, M. Maher, R. O'Kennedy, T. J. Smith, A. J. Riccoa and L. P. Lee, Integrated microfluidic tmRNA purification and real-time NASBA device for molecular diagnostics, *Lab Chip*, **8**, pp. 2071–2078, 2008.

[96] K. S. Drese, O. Soerensen, L. Solli and A. Gulliksen, smallTalk2003, *The Microfluidics, Microarrays and BioMEMS conference*, San Jose, USA, p. 69, 2003.

[97] J. Y. Shin, J. Y. Park, C. Y. Liu, J. S. He, S. C. Kim, Chemical structure and physical properties of cyclic olefin copolymers - (IUPAC technical report), *Pure Appl. Chem.*, Vol. 5, **77**, , 801-814, 2005.

[98] D. D. Blumberg, Creating a ribonuclease-free environment, Methods Enzymol., 152, pp. 20-24, 1987.

[99] J. Mizuno, H. Ishida, S. Farrens, V. Dragoi, H. Shinohara, T. Suzuki, M. Ishizuka, T. Glinsner, F. P. Lindner, S. Shoji, Cyclo-olefin polymer direct bonding using low temperature plasma activation bonding, Transducers, Seoul, Korea, pp. 1346-1349, 2005.

[100] T. B. Christensen, C. M. Pedersen, K. G. Groendahl, T. G. Jensen, A. Sekulovic, D. D. Bang and A. Wolff, PCR biocompatibility of lab-on-a-chip and MEMS materials, *J. Micromech. Microeng.*, **17**, pp. 1527–1532, 2007.

[101] Bio-Disk - A Centrifugal Platform for Integrated Point-of-care

Diagnostics on whole blood, Ziwschenbericht CPI, funded by the Landesstiftung Baden-Wuerttemberg (contract 24-720.431-1-7/2).

[102] O. Prucker et al., Photochemical Attachment of Polymer Films to Solid surfaces via Monolayers of Benzophenone Derivative, *J. Am. Chem. Soc.*, **121**, pp. 8766-8770, 1999.

[103] S. B. Petersen, V. Jonson, P. Fojan, R. Wimmer and S. Pedersen, Sorbitol prevents the self-aggregation of unfolded lysozyme leading to an up to 13 °C stabilisation of the folded form, *Journal of Biotechnology*, Vol 114, **3**, pp. 269-278, 2004.

[104] R. Chakrabarti, C. E. Schutt, The enhancement of PCR amplification by low molecular-weight sulfones, *Gene*, **274**, pp. 293–298, 2001.

[105] Triton Technology, T_g and Melting Point of a Series of Polyethylene Glycols using the Material Pockets, www.triton-technology.co.uk, accessed 2010

[106] M. Focke, F. Stumpf, G. Roth, R. Zengerle and F. v. Stetten, Centrifugal microfluidic system for primary and secondary PCR, to be submitted, 2010.

[107] J. Steigert, S. Haeberle, T. Brenner, C. Mueller, C. P. Steinert, P. Koltay, N. Gottschlich, H. Reinecke, J. Ruehe, R. Zengerle and J. Ducrée, *J. Micromech. Microeng.*, **17**, pp 1–9, 2007.

[108] M. M. Dudek, R. P. Gandhiraman, C. Volcke, A. A. Cafolla, S. Daniels and A. J. Killard, Plasma surface modification of cyclo-olefin polymers and its application to lateral flow bioassays, *Langmuir*, **25**, pp. 11155–61, 2009.

[109] C. Ahn, S. Kim, H. Cho, S. Murugesan and G. Beaucage G, Surface modification of cyclic olefin copolymers for bio-MEMS microfluidic devices, *MRS Symp. Proc.*, San Francisco, pp. 1101–4, 2002.

[110] Y. Liu, X. Chen and J H Xin, Superhydrophobic surfaces from a simple coating method: a bionic nanoengineering approach, 2006, *Nanotechnology*, **17**, pp 3259–3263.

[111] A. L. Tiensuu, O. Öhman, O. Lundbladh, O. Larsson, Hydrophobic valves by ink-jet printing on plastic CDs with integrated microfluidics, *Proc. MicroTAS*, pp 575–578, 2000.

[112] O. Larsson, A.-L. Tiensuu, Hydrophobic Barriers, United States Patent Application US2007/0059216 A1, 2007.

[113] Teflon AF, Processing and Use, Product Information, Du Pont, USA, www.dupont.com, accessed 2010.

7.1 Literature

[114] A.I. Caçoa, A. M. A. Diasa, I. M. Marruchoa, M. M. Piñeirob, Luis B. Santosc, J. A. P. Coutinhoa, Thermophysical properties of some perfluorocompounds, *Proc. 15th Symposium on Thermophysical Properties*, Boulder, USA, 2003.

[115] K. Watanabe and M. Okada, Measurements of the Surface Tension of Four Halogenated Hydrocarbons, CCI3F, CCI2F, CCI3F3, and C2CI2F4, *International Journal of Thermophysics*, Vol. 2, **2**, 1981.

[116] Masahiko Yasumoto, Yasufu Yamada, Jyunji Murata, Shingo Urata, and Katsuto Otake, Critical Parameters and Vapor Pressure Measurements of Hydrofluoroethers at High Temperatures, *J. Chem. Eng. Data*, **48**, 1368-1379, 2003.

[117] 3M product information, Fluorinert Electronic Liquid FC-77, FC-70, FC-40, Novec Engineering Fluid HFE-7200, www.mmm.com, accessed 2010.

[118] S. Schrader, D. Prescher and V. Zauls, New chromophores and polymers for second-order nonlinear optics, *Proc. SPIE*, 3474, pp 160-171, 1998.

[119] S. Schrader, R. Wortmann, D. Prescher, K. Lukaszuk and A. H. Otto, New chromophores and Solvatochromy and electro-optical study of new fluorine-containing chromophorespolymers for second-order nonlinear optics, *Proc. SPIE*, 3474, pp 14-22, 1998.

[120] John R. Moffatt, Tim A. Beerling, David A. Neel, Use of perfluorinated compounds as a vehicle component in ink-jet inks, United States Patent 5919293, 1999.

[121] A. P. Alivisatos, Semiconductor clusters, nanocrystals, and quantum dots, *Science*, **271**, pp 933 937, 1996.

[122] R. J. Klein, P. M. Biesheuvel, B. C. Yu, C. D. Meinhart, and F. F. Lange, Producing Super-Hydrophobic Surfaces with Nano-silica Spheres, *Z. Metallkd.*, **94**, pp 377–380, 2003.

[123] A. Gulliksen, L. Solli, F. Karlsen, H. Rogne, E. Hovig, T. Nordstrøm and R. Sirevåg, Real-Time Nucleic Acid Sequence-Based Amplification in Nanoliter Volumes, *Anal. Chem.*, **76**, pp. 9-14, 2004.

[124] A. Borst, A. Box and A. Fluit, False-positive results and contamination in nucleic acid amplification assays: Suggestions for a prevent and destroy strategy, *Eur J Clin Microbiol Infect Dis*, Vol. 23, **4**, pp. 289 - 299, 2004.

[125] Project report, D5.2, Microactive, www.sintef.no/Projectweb/ Microactive, accessed 2010.

[126] T. Laurell, L. Wallman, J. Nilsson, Design and development of a silicon microfabricated flow-through cell for on-line picoliter sample handling. *J. Micromech. Microeng.*, **9**, pp. 369-376, 1999.

[127] W. Streule, T. Lindemann, G. Birkle, R. Zengerle, and P. Koltay, PipeJet: A Simple Disposable Dispenser for the Nano- and Microliter Range, *JALA*, Vol. 9, **5**, pp.300-306, 2004.

[128] K. S. Birdi, D. T. Vu and A. Winter, A Study of the Evaporation Rates of Small Water Drops Placed on a Solid Surface, *J. Phys. Chem.*, **93**, pp. 3102-3703, 1989.

[129] Martin Christ seminar, Freeze-drying of pharmaceutics, presentation slides, 2003.

[130] United States Patent #5556771, 1996.

[131] H.Gieseler, and G. Lee, Influence of Different Cooling Rates on Cake Structure of Freeze-dried Samples Measured by Microbalance Technique". Poster presentation, Controlled Release Society, German Chapter, Annual Meeting, Munich, 2003.

[132] World Patent Application, WO 95/27721, 1995.

[133] United States Patent #7256000, 2007.

[134] United States Patent #3928566, 1975.

[135] G. Rowley, Quantifying electrostatic interactions in pharmaceutical solid systems, *International Journal of Pharmaceutics*, **227**, pp. 47–55, 2001.

[136] Y.-H. Chang, G.-B. Lee, F.-C. Huang, Y.-Y. Chen, J.-L. Lin, Integrated polymerase chain reaction chips utilizing digital microfluidics, *Biomed Microdevices*, **8**, pp. 215–225, 2006.

[137] J. Black, Biological performance of materials: fundamentals of biocompatibility, CRC Press, USA, 2005.

[138] S. Garst, M. Schuenemann, M. Solomon, M. Atkin, E. Harvey, Fabrication of multilayered microfluidic 3D polymer packages, 55 th Electronic Components and Technology Conference, May 31 - June 3, Lake Buena Vista, United States, 1, pp. 603-610, 2005.

[139] Cycloolefin Copolymer (COC) Topas®, Product brochure, TOPAS Advanced Polymers GmbH, Germany, www.topas.com, accessed 2010.

[140] J. A. Forrest, K. Dalnoki-Veress and J. R. Dutcher, Interface and chain confinement effects on the glass transition temperature of thin polymer films, *Physical Review E*, Vol.56, **5**, pp. 5705-5715, 1997.

7.1 Literature

[141] A. Piruska, I. Nikcevic, S. H. Lee, C. Ahn, W. R. Heineman, P. A. Limbach, C. J. Seliskar, The autofluorescence of plastic materials and chips measured under laser irradiation, *Lab Chip,* 5(12), pp. 1348-1354, 2005.

[142] A. G. Emslie, F. T. Bonner and L. G. Peck, Flow of a Viscous Liquid on a Rotating Disk, *Journal of Applied Physics*, Vol. 29, **5**, pp. 858-862, 1958.

[143] A. Puntambekar et al., Effects on surface modification on thermoplastic fusion bonding for 3-D microfluidics, *Proceeding µTas*, pp. 425-427, 2002.

[144] T. Nielsen, M. Vogler, F. Reuter, G. Gruetzner, A. Kristensen, Dissolution Investigations of Topas for Homogeneous Imprints, *Proc. NNT*, 2004.

[145] Environmental Health Criteria for Toluene, IPCS, Inchem, www.inchem.org, accessed 2010.

[146] C. J. Lawrence, The mechanics of spin coating of polymer films, *Phys. Fluids*, Vol. 10, **31**, pp. 2786-2795, 1988.

[147] F. Bundgaard, T. Nielsen, D. Nilsson, P. Shi, G. Perozziello, A. Kristensen, O. Ceschek, Cyclic Olefin Copolymers - An exceptional material for exceptional Lab-on-a-chip systems, *Proc. MicroTas*, pp. 372-374, 2004.

[148] G. Carrcano, M. Ceriani, and F. Soglio, Spin Coating with High Viscosity Photoresist on Square Substrates, *Hybrid Circuits*, Vol. 32, p. 12, 1993.

[149] Lab for process technology, IMTEK, University of Freiburg, Germany, www.imtek.de/prozesst, accessed 2010.

[150] S. J. Pomfret, K. P. Plucknett, V. Normand, W. J. Frith, I. T. Norton, Interfacial Adhesion of Biopolymer Gels Measured using the Peel Test, Mat. Res. Soc. Symp., Vol. 622, FF7.3.1-6, 2000.

[151] O. Cohu, A. Magnin, Forward roll coating of Newtonian fluids with deformable rolls: an experimental investigation, *Chem. Eng. Sc.*, **52**, pp. 1339-1347, 1997.

[152] Naturwissenschaftliches und Medizinisches Institut an der Universität Tübingen (NMI), Germany, www.nmi.de, accessed 2010.

[153] J. Y. Shin, J. Y. Park, C. Liu, J. He, S. C. Kim, Chemical structure and physical Properties of Cyclic olefin Copolymers, *Pure Appl. Chem.*, Vol. 77, **5**, pp. 801-814, *2005*.

[154] United States Pharmacopeia, www.usp.org, accessed 2010.
[155] United States Patent #5262475, 1993.
[156] S. Geng, S. Zhang and H. Onishi, XPS Applications in Thin Films Research, *Materials Technology*, Vol. 17, **4**, pp. 234-249, 2002.
[157] R. W. Paynter, XPS studies of the ageing of plasma-treated polymer surfaces, *Surf. Interface Anal.*, **29**, pp. 56–64, 2000.

7.2 Used Materials

[M1] Cycloolefin-Copolymer (COC) Topas 5013, Kunststoff-Zentrum Leipzig gGmbH, Germany, www.kuz-leipzig.de, accessed 2010.
[M2] Water, Molecular Biology Reagent, W4502, Sigma-Aldrich, www.sigma.com, accessed 2010.
[M3] RNaseAlert kit, Ambion, www.ambion.com, USA, accessed 2010.
[M4] P2263, Polyethyleneglycol Bisphenol a Epichlorohydrin Copolymer, Sigma-Aldrich, www.sigma-aldrich.com, accessed 2010.
[M5] Teflon AF 1600, DuPont, USA, www.dupont.com, accessed 2010.
[M6] Vistex II Anti-Fog, Film Specialities Incorporated, USA, www.film-specialties.com, accessed 2010.
[M7] Fluorinert FC-77, 3M Corporation, USA, www.mmm.com, accessed 2010.
[M8] Lumidot 640, Sigma-Aldrich, USA, www.sigma-aldrich.com, accessed 2010.
[M9] Lumidots, Sigma-Aldrich, USA, www.sigma-aldrich.com, accessed 2010.
[M10] CFC-11, Fluorotrichloromethane, USA, Sigma-Aldrich, www.sigma-aldrich.com, accessed 2010.
[M11] Ruß, Typ 901, Degussa / Evonik, Germany, corporate.evonik.com, accessed 2010.
[M12] Polyolefin foil 9795R, 3M Corporation, USA, www.mmm.com, accessed 2010.
[M13] Safe-Lock Microcentrifuge Tube, Eppendorf, Germany, www.eppendorf.com, accessed 2010.
[M14] P5413, Poly(ethylene glycol), Sigma-Aldrich, www.sigma-aldrich.com, accessed 2010.
[M15] ZeonorFilm ZF-14-188, Zeonor Corporation, Japan, www.zeorex.com, accessed 2010.

[M16] Cycloolefin-Copolymer (COC) Topas 8007, Kunststoff-Zentrum Leipzig gGmbH, Germany, www.kuz-leipzig.de, accessed 2010.

[M17] E480R, Zeonor Corporation, Japan, www.zeonex.com, accessed 2010.

[M18] Epo-Tek 375, Epoxy Technology, United States, www.epotek.com, accessed 2010.

[M19] Epo-Tek 302-3M, Epoxy Technology, United States, www.epotek.com, accessed 2010.

[M20] Bessey TPN-BE, Bessey, Germany, www.bessey.de, accessed 2010.

7.3 Used Devices

[D1] CKX 41, Olympus Deutschland GmbH, Germany, www.olympus.de, 2010.

[D2] Zeiss AxioCam MRc, Carl Zeiss MicroImaging GmbH, Germany, www.zeiss.de, accessed 2010.

[D3] PICCOLO GHz, Plasma Electronic GmbH, Germany, www.plasma-electronic.de (accessed 2010).

[D4] PipeJet P18, BioFluidiX GmbH, Germany, www.biofluidix.de, accessed 2010.

[D5] Zwicki I Z2.5/TN1S, Zwick GmbH & Co. KG, Germany, www.zwick.de (accessed 2010).

[D6] OCA 15+, DataPhysics Instruments GmbH, Germany, www.dataphysics.de (accessed 2010).

[D7] AFM, Nanoscope III, Digital Instruments / VEECO, USA, www.di.com, accessed 2010.

[D8] Zygo New View 5000, USA, www.zygo.com, accessed 2010.

[D9] REM, Jeol, Germany, www.jeol.de, accessed 2010

[D10] Axiophot, Carl Zeiss MicroImaging GmbH, Germany, www.zeiss.de, accessed 2010.

[D11] PipeJet P9, BioFluidiX GmbH, Germany, www.biofluidix.de, accessed 2010.

[D12] BioSpot 160, BioFluidiX GmbH, Germany, www.biofluidix.de, accessed 2010.

[D13] PipeJet Tip I, BioFluidiX GmbH, Germany, www.biofluidix.de, accessed 2010.

[D14] microbalance SC2, Satorius AG, Germany, www.satorius.de (accessed 2010).

7.3 Used Devices

[D15] Luxeon III Star, blue, Lumileds / Phillips, The Netherlands, www.philipslumileds.com, accessed 2010.

[D16] GFP Filter Set, Semrock, USA, www.semrock.com, accessed 2010.

[D17] Alpha 2-4 LSC, Martin Christ GmbH, Germany, www.martinchrist.de, accessed 2010.

[D18] Tencor P10, KLA Tencor, USA, www.kla-tencor.com, accessed 2010.

[D19] Stork Laminator, Stork Vertrieb KG, Germany, storkvertrieb@aol.com.

[D20] SA1-K, Self-Adhesive Thermocouple, Omega, Germany, www.omega.com, accessed 2010.

[D21] Fluke 54 II Thermometer, Fluke, USA, www.fluke.com, accessed 2010.

[D22] Tempilstick 65°C, Tempil, USA, www.tempil.com, accessed 2010.

[D23] Binder BF Series, Binder, Germany, www.binder-world.com, accessed 2010.

[D24] XPS 5600 ci, Physical Electronics, USA, www.phi.com, accessed 2010.

Chapter 8
Appendix

8.1 Process Recommendation

8.1.1 Low Temperature Applications

Step	Description	Parameter	Type
1	Rinsing	* Put chips in 2-propanol bath * Ultrasonicate, 30min @ 100 % * Dry with pressurized nitrogen	Surface Modification
2	Sterilization	* Put chips in 3 % hydrogen peroxide bath * Store chips for 24h * Rinse with RNase-free water [M2] * Dry with pressurized nitrogen	Surface Modification
3	Surface Activation	* Evacuate plasma chamber $p = 1$ Pa * Ingest oxygen $p = 10$ Pa * Microwave on [D3] 120 s, 200 W	Surface Modification

4	Hydrophilic Coating	* Pipette:
		  - 5 wt% PEG [M4]
		  - methanol
		  into channels (100 nL/mm^2 coverage)
		* Dry for 120 s
		* Remove spill with
		  2-propanol-soaked cloth
			Surface Modification

5	Hydrophobic Patterning	* Dispense 10 nL droplets [D11] of:
		  - 0.5 wt% Teflon [M5]
		  - 0.25 wt% carbon black [M11]
		  - FC-77 [M7]
		  into restrictions (150 nL/mm^2 coverage)
			Surface Modification

6	Spotting & Drying	* Dispense 25 nL droplets [D11,D13] of:
		  - primers
		  - molecular beacon probes
		  - RNase-free water [M2]
		  into reaction chambers
		* Dry for 60 s
			Reagent Storage

8.1 Process Recommendation

7	Spotting & Freezing	* Store chips and chip holder for 2h at -80°C
		* Transport chips in dry ice
		* Dispense 50 nL droplets [D11,D13] of:
		  - enzymes
		  - lyoprotectants
		  - cryoprotectants
		  - RNase-free water [M2]
		  into reaction chambers on frozen chip
		* Put chips in freeze-dryer [D17] cavity
			Reagent Storage

8	Freeze-Drying	* Conduct primary drying for 20 min
		  - Evacuate freeze-dryer to 10 Pa
		  - Freeze-dryer cavity temperature -35°C
		* Conduct primary drying for 30 min
		  - Heat up freeze-dryer cavity to -10°C
		* Conduct primary drying for 45 min
		* Conduct secondary drying for 10 min
		  - Evacuate freeze-dryer to 2 Pa
		* Conduct secondary drying for 30 min
		  - Heat up freeze-dryer cavity to 10°C
		* Conduct secondary drying for 30 min
		* Conduct secondary drying for 120 min
		  - Heat up freeze-dryer cavity to 25°C
		* Put chips in protective container, remove chips from freeze-dryer cavity, store
			Reagent Storage

9	Temperature Diffusion Bonding	* Align chip & compound foil & protective foil in chip holder
		* Move assembly through laminator
		  - $v_{roll} = 0.4$ m/min
		  - $T_{roll} = 140$°C
		  - $p_{roll} = 2$ bar
		  repeat procedure once
			Biocompatible Sealing

8.1.2 High Temperature Applications

Step	Description	Parameter	Type
1	Rinsing	* Put chips in 2-propanol bath * Ultrasonicate, 30min @ 100 % * Dry with pressurized nitrogen	Surface Modification
2	Sterilization	* Put chips in 3 % hydrogen peroxide bath * Store chips for 24h * Rinse with RNase-free water [M2] * Dry with pressurized nitrogen	Surface Modification
3	Surface Activation	* Evacuate plasma chamber $p = 1$ Pa * Ingest oxygen $p = 10$ Pa * Microwave on [D3] 120 s, 200 W	Surface Modification

4	Hydrophilic Coating	* Pipette:
		  - 2 vol% Vistex [M6]
		  - 2-propanol
		  into channels (50 nL/mm^2 coverage)
		* Cure chips for 45 min @ 120°C
			Surface Modification

5	Hydrophobic Patterning	* Dispense 10 nL droplets [D11] of:
		  - 0.5 wt% Teflon [M5]
		  - 0.25 wt% carbon black [M11]
		  - FC-77 [M7]
		  into restrictions (150 nL/mm^2 coverage)
			Surface Modification

6	Spotting & Drying	* Dispense 25 nL droplets [D11,D13] of:
		  - primers
		  - molecular beacon probes
		  - RNase-free water [M2]
		  into reaction chambers
		* Dry for 60 s
			Reagent Storage

7	Adhesive Bonding	* Transfer 2.5 μm of adhesive [M18] onto chip surface * Align chip & lid in chip holder * Apply force > 1 kN on chip assemblies * Evacuate assembly for 3h @ 10 mbar * Cure assembly for 3h @ 70°C	Biocompatible Sealing

Chapter 9
Nomenclature

9.1 List of Symbols

Symbol	Description	Unit
A	Surface area	m^2
dA	Surface element	m^2
CV	Coefficient of variation	%
d	diameter	m
d_{adh}	Adhesive layer thickness	µm
$d_{channels}$	Distance between channels	m
η	Dynamic viscosity	Pa s
η_{adh}	Adhesive viscosity	Pa s
F	Force	N
F_s	Intermolecular force	N
h	Channel height	m
l	Length	m
M_w	Average molecular weight	g/mol
p	Pressure	Pa
p_θ	Capillary pressure	Pa
Δp_θ	Difference in capillary pressure	Pa
p_{roll}	Roll pressure	bar
θ	Static contact angle	°
$\theta_{channel}$	Static contact angle of channel surface	°
θ_{lid}	Static contact angle of lid surface	°
r	Radius of capillary	m
σ	Surface tension	N/m
σ_{lg}	Surface tension of liquid-gas interface	N/m

σ_{sg}	Surface tension of solid-gas interface		N/m
σ_{sl}	Surface tension of solid-liquid interface		N/m
t	Time		s
t_{eva}	Evaporation time		s
t_{store}	Storage time		s
t_{vac}	Evacuation time		s
τ	Shear rate		1/s
T	Temperature		°C
T_g	Glass transition temperature		°C
T_{roll}	Roll temperature		°C
u	Velocity		m/s
v_{min}	Minimum roll velocity		m/min
v_{roll}	Roll velocity		m/min
w	Channel width		m
dW	Work element		N m
y	Plate distance		m

9.2 List of Abbreviations

Abbreviation	Description
mL	milliliter (10^{-6} m^3 = 1 mL)
µL	microliter (10^{-9} m^3 = 1 µL)
nL	nanoliter (10^{-12} m^3 = 1 nL)
pL	picoliter (10^{-15} m^3 = 1 pL)
AF	Amorphous fluoropolymer
AFM	Atomic force microscopy
a.u.	Arbitrary unit
CaSki	Cervical carcinoma cells
CB	Carbon black

9.2 List of Abbreviations

CCD	Charge-coupled device
CD	Compact disk
CIN	Cervical intrapithelial neoplasia
CNC	Computer numerical control
COC	Cyclic olefin copolymer
COP	Cyclic olefin polymer
DI-water	Deionized water
DMSO	Dimethylsulfoxide
DNA	Deoxyribonucleic acid
EU	European Union
FAM	Fluorescein amidite
FITC	Fluorescein-isothiocyanate
HC2	Hybrid capture 2
HCl	Hydrocholic acid
H_2O_2	Hydrogen peroxide
HPV	Human papillomavirus
IMTEK	Department of Microsystems Engineering
KCl	Potassium chloride
LED	Light-emitting diode
LoaC	Lab-on-a-Chip
MEMS	Micro-electro-mechanical systems
$MgCl_2$	Magnesium chloride
mRNA	Messenger RNA
μTAS	Micro total analysis system
NASBA	Nucleic acid sequence-based amplification
O_2	Oxygen
PCR	Polymerase chain reaction
PDMAA-BP	Poyl-dimethyl-acrylamid-benzophenone

9.2 List of Abbreviations

PEG	Poly-ethylene-glycole
PEtOx-BP	Poly-ethylene-oxazoline-benzophenone
QD	Quantum dots
RNA	Ribonucleic acid
RNase	Ribonuclease
SEM	Scanning electron microscope
SNR	Signal-to-noise ratio
Teflon	Poly-tetrafluorethylene, PTFE
USP	United States Pharmacopeia
UV	Ultraviolet
VaIN	Vaginal intrapithelial neoplasia
VAIN	Vulval intrapithelial neoplasia
XPS	X-ray photoelectron spectroscopy

I want morebooks!

Buy your books fast and straightforward online - at one of world's fastest growing online book stores! Environmentally sound due to Print-on-Demand technologies.

Buy your books online at
www.morebooks.shop

Kaufen Sie Ihre Bücher schnell und unkompliziert online – auf einer der am schnellsten wachsenden Buchhandelsplattformen weltweit! Dank Print-On-Demand umwelt- und ressourcenschonend produziert.

Bücher schneller online kaufen
www.morebooks.shop

KS OmniScriptum Publishing
Brivibas gatve 197
LV-1039 Riga, Latvia
Telefax +371 686 204 55

info@omniscriptum.com
www.omniscriptum.com

Printed by Books on Demand GmbH, Norderstedt / Germany